福特基金会资助项目

中国西部草原可持续发展管理研究
——以黑河流域为例

张利华　王红兵　赵传燕　邵丹娜　著

科学出版社

北　京

图书在版编目（CIP）数据

中国西部草原可持续发展管理研究：以黑河流域为例 / 张利华等著.
—北京：科学出版社. 2016.12

ISBN 978-7-03-050660-3

Ⅰ.①中… Ⅱ.①张… Ⅲ.①草原管理-可持续性发展-研究-中国 Ⅳ.①S812.5

中国版本图书馆CIP数据核字（2016）第274274号

责任编辑：侯俊琳 杨婵娟 张翠霞 / 责任校对：彭 涛

责任印制：徐晓晨 / 整体设计：铭轩堂

编辑部电话：010-64035853

E-mail：houjunlin@mail.sciencep.com

科学出版社出版

北京东黄城根北街16号

邮政编码：100717

http://www.sciencep.com

北京厚诚则铭印刷科技有限公司 印刷

科学出版社发行 各地新华书店经销

*

2016年12月第 一 版 开本：720×1000 1/16

2018年1月第二次印刷 印张：18 1/2 插页：3

字数：352 000

定价：98.00元

（如有印装质量问题，我社负责调换）

序　言

党的十八大以来，以习近平同志为核心的党中央协调推进“五位一体”总体布局和“四个全面”战略布局，牢固树立和贯彻落实创新、协调、绿色、开放、共享的“五大发展理念”，将生态文明建设摆在更加重要的战略位置，以构建人与自然和谐发展的现代化建设新格局。2015年，中共中央、国务院印发的《关于加快推进生态文明建设的意见》中指出，从总体上看我国生态文明建设水平仍滞后于经济社会发展，资源约束趋紧，环境污染严重，生态系统退化，发展与人口、资源、环境之间的矛盾日益突出，这些已成为制约经济社会可持续发展的重大瓶颈。因此，生态脆弱区的可持续发展管理是我国生态文明建设战略的一个重要组成部分。

草原在我国生态环境保护和经济社会发展中具有重要战略地位。加强草原生态保护、促进牧民增收，对于保障国家生态安全、加快牧区经济社会发展、构建和谐社会具有重大意义。国务院第128次常务会议决定，从2011年开始，国家在内蒙古等8个主要草原牧区省份全面实施草原生态保护补助奖励政策，2012年又将政策实施范围扩大到黑龙江等5个非主要牧区省份的36个牧区、半牧区县，覆盖了全国268个牧区、半牧区县。“十二五”期间，国家启动实施草原补助奖励政策，取得了显著成效，有力促进了牧区草原生态、牧业生产和牧民生活的改善。

本书以西部草原为研究对象，选取黑河流域草原作为核心案例，主要对生态脆弱区草原生态可持续发展管理实践进行分析、探索并提出相应的政策意见和建议。研究包括以下核心内容：①从国家层面和黑河流域区域层面出发，梳理我国草原管理政策体系；②整理和分析黑河流域主要的自然生境要素数据，分析黑河流域草原生境要素的空间变化及草地空间变化；③分别基于农牧民、基层管理者进行问卷调查，根据调查结果，分析黑河

流域草原管理实践；④结合以上三部分分析的主要结论，总结提出西部草原可持续发展管理面临的问题和应对策略，并据此提出管理政策设计的建议。总体上，本书聚焦于草原管理的实践，在方法上结合政策调研和梳理、数据模拟分析和问卷调查，为草原管理政策制定和执行提供借鉴，为草原生态保护和可持续发展管理等研究工作提供理论参考。

自2014年开始，历时两年半的时间，笔者带领的研究团队足迹踏遍黑河流域的区县，发放农牧民调查问卷1500余份，管理者调查问卷200余份，多次组织召开不同层次的草原管理干部与农牧民座谈和访谈会，同时邀请相关领域的专家、学者召开座谈会，修正研究方案，完善研究内容，最终完成了本研究的既定目标。本书在以下两个方面体现了创新成果：一是在研究对象方面，开拓了可持续发展管理实践的研究，已有对可持续发展领域研究成果多集中在目标、评价、系统集成等方面，相对而言管理实践的研究较少；二是在研究方法方面，针对黑河流域草原可持续发展管理政策，分别从农牧民和管理者两个层面进行问卷调查和对比分析，更为全面系统地探究草原可持续发展管理方面的问题，并提出具有较强针对性的政策建议。

最后，衷心感谢在我们调研过程中黑河流域沿途各市、县、区，如青海省祁连县，甘肃省甘州区、山丹县、民乐县、肃南裕固族自治县、阿克塞哈萨克族自治县、高台县、临泽县、肃北蒙古族自治县，内蒙古自治区额济纳旗，以及其以上各区（县）所属乡镇、草原监理站等各级草原管理干部及广大农牧民给予的大力支持与积极认真的参与；衷心感谢以不同方式为项目研究提供帮助和参与项目研究的领导、专家、学者，感谢博士、硕士研究生所付出的积极努力和贡献，他们其中主要有谢园、谢正团、杨劼、尹叶红、何革华、刘学敏、王光辉、刘勇等；同时感谢科学出版社的侯俊琳编审、杨婵娟编辑，是他们的努力推动和辛勤劳作，使这本书能够以较好的形式呈现在读者面前。

限于水平，书中一定存在欠缺和不足之处，请广大读者批评指正。

作　者

2016年10月28日

目 录

Research into the Sustainable Development Management of Western Grasslands in China

A case study of the Heihe River Basin

中国西部草原可持续发展管理研究

——以黑河流域为例

中文篇

第 1 章 绪 论

1.1 研究背景

伴随着全球气候变暖及各种自然灾害的频繁发生，生态环境保护已经成为国际社会的共识。草地是人类生存和发展的基本土地资源，草地资源是一种可更新资源。在人类干预以前，原生草原面积占地球陆地面积的 40% ～ 50%。随着畜牧业的发展，全球草地面积不断缩小。合理开发草地资源，有效保护草地资源，既保护生态环境和生物多样性，又满足人类发展和生活需要，这是一个永恒的世界性话题。

我国是仅次于澳大利亚的世界第二草原大国，拥有天然草原近 4 亿 hm^2 ①，约占我国陆地面积的 41.7%，是耕地面积的 3.2 倍、森林面积的 2.5 倍，在我国农田、森林和草原等绿色植被生态系统中占到 63%，是我国面积最大的陆地生态系统。但 20 世纪 90 年代以来，干旱所导致的各类问题日益凸显，草原生态环境问题越来越得到人们的关注与重视。按照高尚玉等（2008）的统计，20 世纪 50 年代我国北方发生大规模强沙尘暴灾害 5 次，60 年代 8 次，70 年代 13 次，80 年代 14 次，90 年代 23 次。沙尘暴是草原沙化带来的最为直接的自然灾害，草原生态环境的破坏还表现为草原的沙化、退化、盐碱化。据农业部统计，目前我国 90% 的可利用天然草原不同程度地发生退化、沙化、盐碱化，而且每年还以 200 万 hm^2 的速度增加。

生态环境的日益恶化尤其是草原生态环境的破坏，不仅带来了沙尘暴

① $1hm^2=10^4m^2$。

的肆虐，同时严重影响草原地区甚至全人类的生产生活，也引起了党和国家的高度重视。党的十八届三中全会审议通过的《中共中央关于全面深化改革若干重大问题的决定》指出，要加快生态文明制度建设，并以纲要的形式明确提出，建设生态文明，必须建立系统完整的生态文明制度体系，实行最严格的源头保护制度、损害赔偿制度、责任追究制度，完善环境治理和生态修复制度，用制度保护生态环境。2014 年中央一号文件对草原生态保护作了明确规定："从 2014 年开始，继续在陡坡耕地、严重沙化耕地、重要水源地实施退耕还林还草。加大天然草原退牧还草工程实施力度，启动南方草地开发利用和草原自然保护区建设工程。"

1.2 研究意义

西部地区包括内蒙古、新疆、陕西、宁夏、甘肃、青海、四川、重庆、西藏、贵州、云南、广西 12 个省份，土地总面积 6.75 亿 hm^2，其中草原面积 3.31 亿 hm^2，约占土地总面积的 49.04%，约占全国草原总面积的 84.10%（张苏琼和阎万贵，2006）。在草原治理和保护区域划分的问题上，现有的文献（苏大学，2001）将西部草地划分为蒙宁甘温带干草原区、西北干旱荒漠和山地草原区、青藏高原高寒草甸高寒草原区三个区域。按畜牧带划分，可将草地资源划分为内蒙古畜牧带、新疆畜牧带、青海畜牧带及西藏畜牧带。

从系统的观点来看，草原生态系统是一个由自然（包括自然与环境）、经济、社会（包括人口）三个子系统共同构成的复合生态系统。草原在陆地生态系统中的巨大作用并没有被人们所充分认识和重视，在开发、利用草原资源的过程中，只顾眼前的经济利益，违背自然生态规律，人们面临着日渐加深的草原资源危机、日益严重的土地荒漠化、频繁发生的沙尘暴等令人担忧的问题。西部地区大多是风沙区、水土流失区和水源涵养区，草原既是西部地区生态环境系统的主体组成部分，又是少数民族赖以生存和发展的基础。加强西部草原生态环境保护与建设，不仅是国家生态环境建设的根本所在，也是促进西部地区经济社会发展的客观要求。

西部草原地区是我国国民经济和生态环境保育极其敏感和至关重要的

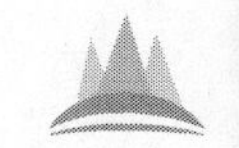

前沿阵地，在国民经济发展及生态安全方面具有不可替代的重要战略地位。然而，近年来，我国“三牧问题”（牧区、牧业、牧民）也比较突出，表现为：草原生态持续退化，牧业成本不断增加，牧民生活水平明显下降，牧区贫富差距正在加剧。导致西部草原生态问题的原因有以下几点：①受经济利益驱动，西部草原载畜量超载过牧严重，承包到户的放牧模式，加之没有适当的划区轮牧措施，每块草原都被频繁利用，牧草毫无生长养息之机，而适口性差和有毒有害植物（如狼毒、丛生灌木等）大量侵入，加剧了草地的退化。②毁林开垦、开矿及陡坡地开垦使得原始植被锐减，水土流失严重；在草原区大量挖掘药材也极易引起草地退化。③西部草地处于内陆干旱半干旱地带，受海洋性气候影响较弱；年降水量在 50 ～ 450 mm，地下水多在 50 ～ 100 mm；乱砍滥伐，使水源涵养林减少，地表及地下水存储量锐减，降水量减少，雪线上升，冰川后退；干旱引起草地加速退化（郭健和郎侠，2005）。④乱捕滥猎，野生动植物丰富区面积不断减少。栖息地环境恶化，造成生物多样性的破坏，导致草原生态平衡被破坏，草场退化，生产率下降。

黑河流域是我国西北干旱区第二大内陆河流域，从发源地到居延海全长 821 km，横跨三种不同的自然环境单元，流域总面积达 1.429×10^5 km^2，位于 96º42′～ 102°00′E、37º41′～ 42°42′N，地跨青海、甘肃和内蒙古 3 省（区）的 11 个县（市、旗），北部与蒙古接壤，东以大黄山与武威盆地相连，西部以黑山与疏勒河流域毗邻。分属三省（区），上游属青海省祁连县，中游属甘肃山丹、民乐、张掖、临泽、高台、肃南、酒泉等市县，下游属甘肃金塔和内蒙古自治区额济纳旗。

该生态系统的特点与西北干旱区的其他内陆河流域具有相似性，从高山冰川（永久积雪）、森林、草地到平原绿洲和戈壁荒漠，构成了一个完整的干旱区自然生态景观。近 50 年以来，黑河流域大兴水利（目前全流域已兴修大小水库 98 座），开垦荒地，形成了大面积的人工绿洲，创造了巨大的农业经济效益。然而，人类活动加剧，导致流域的土地利用和土地覆盖格局发生了深刻的变化，生态环境持续退化。上游地区主要出现了以草原秃斑地和草地沙化、杂毒草蔓延为主要标志的草地退化，天然林减少，生物物种数量减少，以及冰川面积减小、冰川末端波动后退；中游地区主要是土地荒漠化与土壤次生盐碱化；下游地区集中表现为终端湖泊消失，众多天然河道废弃并形成绿洲内部沙源，天然绿洲萎缩，土地沙漠化持续扩张（张志强等，2001）。

第2章 草原管理研究综述

2.1 草地功能研究

2.1.1 生态功能

草地的生态功能是指其生境、生物学性质或生态系统过程（Straton，2006），主要发生在草丛-地境界面，为生命系统提供自然环境条件，具有生命支持功能和环境调节功能，是维持社会与经济发展的基础，主要包括水源涵养、土壤形成、侵蚀控制、废物处理、滞留沙尘和维持生物多样性等功能（柳小妮等，2008）。这些功能是草地生态系统所固有的难以商品化的功能，反映了草地的社会属性，表现为间接的经济价值。

西部草原地区是我国国民经济和生态环境保育极其敏感和至关重要的前沿阵地，占我国国土面积的41.7%，在食物来源和结构改变及生态安全方面具有不可替代的重要战略地位。然而，我国西部草地生态环境脆弱、自然条件恶劣、水肥资源不足，加上乱砍滥伐、滥挖药材、毁林开矿等人为破坏，以及陡坡垦荒、滥垦草场、超载放牧、过度樵采等不合理的开发利用，导致林草植被遭到破坏、生态功能衰退。近年来草地水土流失加剧，退化、沙化严重，草地生态环境在加速恶化，生产力极度下降，单位水平仅相当于美国的1/27（表2-1）。

表2-1　西部草原生态功能区分布与主要生态问题

项目	水源涵养功能区	防风固沙功能区
西部区域分布	甘肃省祁连山高寒荒漠、草原、草甸及其甘南草原，青海省青海湖湿地及上游、黄河源、三江源地区，西藏雅鲁藏布江上游、中喜马拉雅山北翼高寒草原，新疆额尔齐斯－乌伦古河荒漠草原、准噶尔盆地西部山地草原、天山北坡草原与南坡荒漠草原等水源涵养区	陇中－宁中荒漠草原、腾格里沙漠草原荒漠、黑河中下游草原荒漠、毛乌素沙地、柴达木盆地东北部山地高寒荒漠草原、共和盆地草原、阿尔金草原荒漠、准噶尔盆地东部灌木等防风固沙区
主要生态问题	冰川后退，雪线上升；湿地萎缩，植被破坏，草甸退化、沙化；土地荒漠化，土壤保持和水源涵养功能下降；生物多样性维护功能下降	过度放牧、草地过度开垦；水资源不合理开发和过度利用导致草原生态系统严重退化；草地群落结构简单化；物种成分减少；土地沙化面积大、鼠虫害频发；草地生物量和生产力下降；沙尘暴频繁

资料来源：王世金等（2010）

2.1.2　生产功能

草地的生产功能是为生命系统生产各种消费资源，是对草地生态系统生产属性的具体反映，主要在草地-动物界面完成，包括养分循环与贮存、固定 CO_2、释放 O_2 和消减 SO_2 等，这些功能是可以商品化的功能，表现为直接经济价值（王启兰等，2008）。

中国约有 4 亿 hm^2 草原，草原具有强大的碳汇功能。现有的研究成果主要集中于 CO_2、温度及湿度等气象因子和草原开垦、放牧、割草、施肥、人工草地建植等草地管理措施对草原碳汇功能的影响及其作用机制等方面，对我国的草原管理实践具有一定的借鉴意义。

1. 关于草原碳汇

世界草原约占全球陆地面积的 1/3，全球草原生态系统中总碳储量约为 308 Pg，其中约 92%（282 Pg）贮存在土壤中，低熵生物量中的碳所占的比例不到 10%（26 Pg），草原土壤有机碳估计占全世界有机碳的 10% ～ 30%（Schuman et al.，2002）。根据农业部《2010 年全国草原监测报告》，中国草原总碳储量约为 42.73 Pg，占世界草原总碳储量的 9% ～ 19%。

相关研究（Piao et al.，2009）指出我国草原碳吸收大于碳排放，20 世纪 80 ～ 90 年代，每年固定（0.0076±0.002）PgC，青藏高原草原碳储量占全国草原总储量的 50% 左右，每年固定约 55.9 gC/m^2。Wang 等（2011）预测到 2020 年，随着约 1.5 亿 hm^2 草原围栏封育和 0.3 亿 hm^2 人工草地建

植计划的实施，我国草原每年可以固定0.24 PgC。Liebig等（2010）依据44年的草原土壤有机碳数据认定北美大平原草原为重要的碳汇，总的碳固持率为每年每公顷0.39～0.46 MgC。其他学者的研究（Lu et al.，2009；Wang et al.，2011）认为，草原碳会发生汇和源的转换，主要受温度、降雨等气候因子和草原管理措施影响。

2. 气候变化对碳汇的影响

任继周等（2011）认为，我国草地的碳汇主体为植被生产力较高的高山草地、温带湿润草地、斯泰普草地和半荒漠草地，而植被生产力较低的冷荒漠尽管面积较大，但是草地固碳潜力较小。另有研究（Bai et al.，2004；Lesica and Kittelson，2010）表明，随着空气中CO_2浓度的增加，草原生产力水平和水分利用效率均有所提高，草原固碳能力增强；在水缺乏的荒漠草原和典型草原，生物量与降水显著相关，降水增多会提高草原生产力，增加草原碳固持能力。干旱胁迫降低了生长季内蒙古锡林河流域羊草草原生态系统生产力和碳累积量，使生态系统由碳汇变成碳源（Hao et al.，2010）。在青藏高原高寒草地，草地的固碳能力随着降水的增加而增加，而与温度关系不大（Yang et al.，2008）。

3. 草原退化对碳汇的影响

相关研究（吴永胜等，2010；赵锦梅，2006）表明，随着退化程度的加深，土壤有机碳的含量明显减少。同时，也有研究表明由于重度退化而造成的突然（0～20 cm）碳丢失量为3.80 kg/m^2，相当于50.87%的有机碳流失（王文颖等，2006）。Wang等（2011）估计中国北方草地从20世纪60年代到90年代在中度退化和重度退化草地共造成了接近1.24 Pg的净碳损失。在干旱地区，由于严酷的自然条件及人类活动的干扰，土壤很容易发生沙化，土壤沙化会引发如侵蚀、盐渍化和板结等退化过程；干旱区土地退化和荒漠化的直接作用是促进了土壤有机碳和无机碳的矿化，增加了碳的排放，间接作用是通过减少地上生物量生产，导致土壤结构的恶化和土壤容重的增加；无机碳在干旱区土壤中的储存量为土壤有机碳储量的2～5倍（樊恒文等，2002）。因此，干旱地区控制土壤无机碳的截留对于碳汇功能的发挥具有重要的意义。

2.1.3 生活功能

除了以上讨论的草原生态功能以外，草地还需要满足特定环境条件下的生存和生产方式，它是人-草-畜-生态有机结合的载体。草原生活功能主要发生在草畜-经营管理界面，它包括畜牧生产、文化传承和休闲旅游等功能，有些功能可以被商品化，表现为直接经济价值，有些不能被量化，则表现为间接经济价值。

“三牧”已经引起了党中央和国务院的高度重视，2010 年 10 月 12 日，时任国务院总理温家宝主持召开国务院常务会议，决定建立草原生态保护补助奖励机制以促进牧民增收。从 2011 年起，中央财政每年安排资金 134 亿元，在内蒙古、新疆（含新疆生产建设兵团）、西藏、青海、四川、甘肃、宁夏和云南 8 个主要草原牧区，全面建立草原生态保护补助奖励机制（杜发春和曹谦，2012）。

西部草地家畜数量的增加和草地生产力的下降，使草畜矛盾更加突出，这种矛盾必然转化为草地超载过牧、草地退化。草地生产力变化显著，丰收年牧草量高时达到正常年份的 155%，而碰到大的自然灾害时则只有正常年份的 75%。西部草地绝大部分处于自然生长状态，受降水、温度等气候条件的限制，年际丰歉不同，尤其是降水量的多少和降水时间前后直接关系到本年甚至来年草地的植物生长量，这也是一个不利于可持续利用的因素。随着草原生态环境的破坏，草地畜牧业变得更加脆弱和不稳定，如青海省每年死亡牲畜 150 万头，家畜生产性能不断下降（闫志坚等，2006）。

由于自然条件和历史原因，西部草原区社会经济和科学文化长期处于贫困落后状态，特别是近几十年草原生态环境进一步恶化，更加剧了该地区贫困人口的增加。1995 年全国各省（自治区、直辖市）单位面积固定资产原值，上海为 2465 万元 /km^2，北京为 635 万元 /km^2，江苏为 160 万元 /km^2，而西部青海为 3.11 万元 /km^2，内蒙古为 5.18 万元 /km^2，甘肃为 15 万元 /km^2。到 1999 年我国西北仍有 153 个旗县未摆脱贫困，牧民的温饱未能得到彻底解决（闫志坚等，2006）。

另外，随着生态移民工程和新农村、新牧区建设的推进，游牧民定居问题的研究近年来也引起了学界的广泛关注，内容涉及政策导向、实施范围及方法、文化变迁、生活适应性等。另外，有关研究（李静等，2011）指出，对民族心理的调试是牧民首先面对的考验，也是游牧民能否实现永久定居的关键，因此，牧民的心理承受能力、民族传统文化保护、定居模式的微观建构、城市化进程中的民族问题及社区民族工作的研究、定居政策及实施情况都是需要进一步关注的问题。

就缓解草原“三牧”而言，现有国家层面的法律法规和相关政策如表2-2所示。

表2-2 缓解草原“三牧”问题的法律法规及相关政策

颁布时间	法律法规	颁布部门	主要内容
2002年	《中华人民共和国草原法》（修订）	全国人大常委会	不得超过载畜量；牧区的草原承包经营者应当实行划区轮牧；国家提倡在农区、半农半牧区和有条件的牧区实行牲畜圈养，逐步改变依赖天然草地放牧的生产方式
2002年	《国务院关于加强草原保护与建设的若干意见》	国务院	建立基本草地保护制度；实行草畜平衡制度；因地制宜推进划区轮牧、休牧、禁牧制度；加强围栏和牧区水利建设；推行舍饲圈养方式
2002年	《关于加强畜牧业发展的意见》	农业部	合理使用草地资源；草原牧区要推行以草定畜，划区轮牧，科学管理，提高草地畜牧业的综合效益；半农半牧区实行草地轮作，舍饲圈养
2005年	《中华人民共和国畜牧法》	全国人大常委会	国家支持草原牧区开展草原围栏、草原水利、草原改良、饲草饲料基地等草原基本建设，优化畜群结构，改良牲畜品种，发展舍饲圈养，划区轮牧，逐步实现畜草平衡，改善草原生态环境
2007年	《关于促进畜牧业持续健康发展的意见》	国务院	全面推行草畜平衡，实施天然草原禁牧休牧轮牧制度，保护天然草场，建设饲草基地，推广舍饲半舍饲饲养技术，增强草原畜牧业的发展能力
2011年	《关于促进牧区又好又快发展的若干意见》	国务院	有步骤地推行草原禁牧休牧轮牧制度，减少天然草原超载牲畜数量，实现草畜平衡。加强草原围栏和棚圈建设，在具备条件的地区稳步开展牧区水利建设、发展节水高效灌溉饲草基地，促进草原畜牧业由天然放牧向舍饲、半舍饲转变，实现禁牧不禁养

资料来源：中国政府网及农业部相关网站

2.2 草地管理理论研究

现有研究将国内草原管理理论分为四大类，即系统发展论、本土发展论、替代发展论和游牧发展论（表2-3）。

表2-3 草原管理的基本理论

相关理论	主要内容	代表人物
系统发展论	四个生产层理论和系统耦合理论，利用界面理论，实现资源的优化和各层次的发展	任继周
本土发展论	强调本土经验，对于当地发展不相适宜的策略持反对和抵制态度，该理论充分尊重草原畜牧业的发展规律，谨慎地对草原农业生态系统施加影响	达林太、王晓毅等
替代发展论	通过改变牧区的生产方式或者让牧区休养生息，将牧区的生产转移到其他区域，实现牧区生产功能被替代，实现草原恢复	张新时、许鹏、蒋高明等
游牧发展论	要恢复草原，保持草原的可持续发展，必须全面恢复游牧，消灭游牧等于消灭游牧文化，就等于毁灭草原，文化荒漠化是草原荒漠化与沙尘暴的根源	松原正毅等

资料来源：段庆伟等（2012）

相对而言，许鹏（2004）和张新时等（2004）在草原恢复和重建策略方面的研究在本质上具有一定的相似性。任继周等（2011）提出的系统耦合理论从根本上也是与许鹏和张新时的观点一致，归根结底是资源整合和开发新的资源，保护好已经破坏的资源。王晓毅（2009）认为，“单纯的关注草原植被，忽视甚至依靠开采地下水恢复草原，结果只会对草原的生态带来更严重的问题”。因此，发展人工草地时，对地下水的破坏会对草原生态系统造成更深远的影响。

草原生态恢复和建设涉及生态学、资源与环境学、经济学等理论，是一个多学科交叉的研究领域。

2.2.1 公共物品理论

微观经济学通常将公共物品定义为具有竞争性和排他性两大属性特征。应用同样的物品分类标准，不同经济学家对物品的分类也不尽相同。例如，曼昆（Mankiv，2003）将物品分为私人物品、公共物品、共有资源及具有自然垄断属性的四类；文森特·奥斯特鲁姆（Vincent Ostrom）和埃莉诺·奥斯特鲁姆（Elinor Ostrom）（2000）将物品分为私人物品、公共物品、收费物品及公共池塘资源；余永定等（1999）则根据非竞争性和非排他性的不同程度，将公共物品分为纯公共物品（pure-public goods）和准公共物品（quasi-public goods）；保罗·萨缪尔森（Paul Anthony Samuelson）认为相对于私人产品而言，公共物品具有消费上的非竞争性和非排他性两个基本特征。

西部草原生态建设提供的草原生态系统服务是一种较为复杂的公共物品。草原生态建设是恢复和保护退化的草原生态系统服务，其最终目的是为了实现和确保草原生态系统服务的可持续供给。根据前人对于公共物品的定义，可以将草原生态系统服务消费上的非竞争性表述为草原生态系统服务供给某个消费者时，并不妨碍他人消费该生态系统服务，即增加一个草原生态系统服务的消费者的边际成本为零。例如，草原生态系统所具有的固碳作用，就减少全球变暖的威胁而言，整个国际社会都将从中受益。另外，草原生态系统服务消费上的非排他性指的是，只向选择付费的那些人提供草原生态系统服务而把其他人排除在外是不可能的。

公共物品的非竞争性和非排他性这两个特征决定了消费和利用草原生态系统服务和其他公共物品一样，都会面临“公地悲剧”和“搭便车”两个问题。如果某种生态资源的所有权没有排他性，那么由市场这只“看不见的手”来配置生态产品等公共物品就会出现“市场失灵”的问题，进而导致公共资源的过度使用，使生态环境和资源遭到破坏，而且这种破坏结果是不可逆转的。因此，就公共物品的供给而言，经济学家和决策管理者们认为公共物品必须依靠“看得见的手”公共产权市场的介入，通过建立一种补偿机制，给予那些为保护生态环境而牺牲自身利益的人们一定的补偿，进而提高全社会福利。另外，公共物品的非竞争性和非排他性并非一成不变，随着产权的明晰、技术的进步及法律法规的健全，公共物品也转换为私人物品，引入市场机制进行交易和管理。

2.2.2 外部性理论

外部性是经济学中一个广泛讨论和应用的话题，也是新旧古典经济学研究的重要范畴。马歇尔在 1890 年所著的《经济学原理》一书中首次提出外部性这个概念，而最先系统论述外部性理论的则是福利经济学的创始人庇古。庇古（2006）对于外部性的本质给出了这样一个描述：“个人 A 在对个人 B 提供某项支付代价的劳动过程中，附带地亦对其他人提供劳务（并非同样的劳务）或损害，而不能从受益的一方取得支付，亦不能对受害的一方施以补偿。”按照福利经济学的观点，外部性是关于某个经济主体对另一个经济主体产生的外部影响，而这样的影响并非通过价格机制来实现的，也就是说不能通过市场价格反映出来，因此社会成本和私人成本之间必然存在差异。从外部性的作用效果来看，外部性可以是本地的，即限定于某个特定区位，也有可能是全球的，它可能是正的外部性（外部经济的生产者私人受益小于社会受益），也有可能是负的外部性（外部不经济的生产者私人边际成本小于社会边际成本）。草原生态系统的建设和恢复，与其提供的生态系统服务具有正的外部性，它不仅能改善区域的生态环境，还有助于实现更大地域范围生态系统的整体性和相关性，在改善整个区域乃至国家的生态环境的同时，实现了人类福祉的提高。

然而，如果对于人类活动导致的生态环境破坏不作为，将导致生态系统的不断恶化，甚至于难以恢复。从外部性理论出发，如果生态环境受益者无须付费，而受损者无法得到补偿，那么生态环境资源配置与利用便很难达到最优状态。因此，需要采取必要的措施来矫正或者消除这种外部性，即外部性内部化。理论上有两种解决方案："科斯定理"和"庇古定理"。前者倾向于个人交易，只要交易成本足够低，以及产权是明确的，个体、社区甚至超国家实体会持续交易他们的权利直到实现环境商品和服务的帕累托最优；后者则倾向于国家制定税收或补贴的政策机制，即通过政府干预的手段，对外部经济行为予以补贴，以补贴外部经济生产者的成本和他们应得的利润，从而增加外部经济的供给，提高整个社会的福利水平。

2.2.3 人地关系理论

人地关系理论是人文地理学的理论基础，着重讨论人类活动与地理环境之间的相互关系。人地关系是指人类活动与地理环境之间的关系，是以地球表面一定地域为基础的人地关系系统，即人与地在特定的地域中通过相互联系、相互作用而形成的一种动态结构系统（吴传钧，2008）。人地关系地域系统理论要求特定区域的人口、资源、环境和经济社会发展之间要保持经常性动态协调关系，简称为 PRED 协调发展。也有不少学者以此为基础讨论可持续发展理论。

对于草原生态恢复和建设，我们需要明确这样一个关系：经济的发展和社会的进步需要通过对自然生态环境系统进行开发利用和索取，但要使人地关系协调发展，人类社会必须适时地调整开发行为，并且对生态环境

的破坏进行补偿，这样才能实现人口、资源、环境和经济社会的协调发展。

2.2.4 可持续发展理论

1987年联合国发表研究报告《我们共同的未来》，其中首次对可持续发展进行了科学的定义：“可持续发展是能够满足当代人需求且不危及后代人满足其自身需要能力的基础上的发展。”这个定义的含义有以下两层意思：①人类的发展需要考虑代际公平问题（生态正义或环境公平），即当代人的经济活动对当代人的经济社会生活造成影响，也会对未来的环境和资源的储备及利用造成影响，应当将可持续发展视为“一个为后代储备总生产能力的计划”；②必须认识到生态环境对于经济社会发展的一种制约作用，所有的经济活动必须在生态许可的范围内进行，以使环境的服务或废物的排放有一定的限制。

经济学家和生态学家对于可持续发展的问题各持不同意见。生态学家注重寻求人与自然的协调平衡与和谐发展，要求人类的发展必须考虑资源环境本底。另外，生态学家将自然资源划分为使用价值和非使用价值，如生态系统具有的净化空气等功能，认为这些自然资源不能被生产资本、人力资本和社会资本等其他形式的资本所替代。经济学家强调的是如何维持和改善人类的生活水平，只要当代人确保他们留给下一代的资本存量不少于当代人的拥有量，自然资源就可以由生产资本、人力资本和社会资本等其他形式的资本予以替代，用于消费和生产。

基于这样的观点分歧，经济学家和生态学家对于资源环境保护和开发的态度也有所不同。生态学家强调事前预防，对于经济学家强调的事后补偿，如“先污染，后治理，先破坏，后恢复”这样的治理思路，生态学家认为是不可行的，因为就自然生态系统的破坏而言，很多时候是不可逆的，自然资本在某些方面具有不可替代性。

2.3 草地管理对策研究

从福利经济学和制度经济学的研究范畴出发，杨理（2010）揭示了目前草原管理面临的困境，即从“公地悲剧”向“围栏陷阱”转变，认为在目前不完善的管理制度环境下，草原退化是理性经济人抢占草原的饲草和水资源等无主租值行为的必然均衡结果，而其解决措施不是鼓励围栏到户，而是多样化的地方社区自主治理。

蒲小鹏等（2011）则从历史沿革的角度梳理了古代天然草原管理思想（表 2-4），并从加强草原执法、加大宣传和增加投资等方面对当前的草原建设提出了建议，包括加强草原保护行政法规的监督和执法力度，通过加强专项经费投入促进草原民族经济发展。

表2-4　中国古代主要草原保护具体措施

游牧部落 / 民族	具体措施
蒙古族	有各种禁忌和处罚规则，不准乱掘、乱挖草地，搬迁时必须把火熄灭，若引起火灾将受到严惩
藏族	对自然的恐惧、敬畏、崇拜等复杂的心理，加之宗教信仰的缘故，在客观上形成了一种人与自然和谐相处的纯朴的自然生态观
党项族（西夏）	西夏的律令中有关畜牧业的法规占有较大的比例，并表现出很重的农牧业属性，国家的一切赏罚唯以农、牧及战争之功过标准来实施，法律为农、牧、战争而立

资料来源：根据蒲小鹏等（2011）整理

从草原生态恢复和建设的角度出发，闫志坚等（2006）提出了具体的技术途径，包括转变畜牧业生产经营方式、以水为中心发展饲草料基地和人工草地、加快草地植被恢复与重建工作等三个方面，同时，也应当从战略上加强草地资源的管理，开展草地资源保护性政策的评价，加大执法力度，实现草地资源管理的领导责任制，通过控制草原牧区人口，提高人们保护草原资源的意识。宋富民和宋鑫焱（2006）提出用机械化恢复和改良中国西部退化天然草原的办法和措施，用机械化恢复和提高退化天然草原牧草的产量和质量，以此进一步缓解和解决中国西部天然草原超载过牧、饲草不足的问题。陈洁和罗丹（2009）将生态草原治理政策存在的问题概括为以下几个方面：①对农牧民的补偿标准偏低，补偿机制不完善；②农牧户的生产资金严重不足，迫切需要信贷支持；③缺乏供应饲草料的长效机制，草畜矛盾依然严峻；④牧业基础设施建设和公共服务供给不足，明显满足不了需求；⑤后续产业发展滞后，农牧民收入受到严重制约。巩芳和常青（2012）根据环境库兹涅茨曲线提出了反映生态环境变化的草原生态环境库兹涅茨曲线，并从理论上分析政府主导型草原生态补偿机制对草原生态环境库兹涅茨曲线的影响，包括横向优化（通过政府主导型草原生态补偿可以使草原生态环境库兹涅茨曲线在收入水平较低时达到顶点，从而加速草原生态系统的修复，实现草原生态经济与社会的发展）和纵向优化（通过降低草原生态环境库兹涅茨曲线顶点的高度，即通过构建完善的政府主导型草原生态补偿机制修复和改善草原生态系统，减少草原生态系统正熵的流入，增加负熵的流入，从而实现草原生态经济的可持续发展）。

相关学者基于“三牧”问题现状，提出了相应的治理对策。周立和董小瑜（2013）讨论了以家庭为基础、以“双权一制”为基本内涵的“草畜双承包”制度导致的“三牧”问题，认为需要通过明晰产权，以及寻求一种以“私地共管”为基础的“合作共管”机制，通过“赋权”“赋能”等市场重新嵌入社会和自然的努力促进牧区的再组织化，并通过采纳“多中心、自主治理”的制度安排缓解存在的问题。在产业发展研究方面，杜发春和曹谦（2012）提出发展生态畜牧业是中国西部地区畜牧业未来经济发展的趋势和出路，认为发展生态畜牧业应按照生产力发展水平、区域特点发展不同的生产模式，而不能采取“一刀切”的发展模式。

在草原适应性管理方面，美国南加州的草地管理方式是一个典型的被动适应性管理的例子，Chadden 等（2004）针对外来物种入侵、过度放牧、气候变化、城镇化等影响，提出了具体的研究计划，还根据文献和数据资料，综合分析得到草原生态系统的现状信息，并给予生态学家、管理者建议和意见，构建系列预测模型，设计计划，建立试验单元。另有澳大利亚东南亚草原生态系统管理则是运用了主动适应性管理的理论和方法，以保护本地植物及其生境的复杂性为目标，基于火烧和放牧利用对不同草地的影响，选择状态–转化模型和决策树等，构建不同干扰方式（放牧、火烧、割草及不干扰）系统变化趋向预测模型，提出具体的试验方案（Wong and Morgan，2007）。侯向阳等（2011）在对我国草原适应性管理的重要任务的基础上，提出加强草原生态系统基础研究、整合多维度理论、知识与方法、开展模式示范、构建数字草原信息网络、提高监测预测能力等重要措施。

在草畜平衡管理制度方面，杨理（2013）认为草畜平衡管理是唯一以法律保障约束牧户之间竞争的协调机制，应在实现原有的草畜动态平衡的基础上，改变草畜平衡管理模式，从使用命令控制型措施调节牲畜和饲草数量平衡向以市场经济措施调节牲畜总量的新草畜平衡管理转变。李艳波和李文军（2012）认为，在资源时空异质性异常显著的干旱半干旱草场，目前将草畜平衡作为落实到每一微观牧户的实施手段，其可行性值得商榷，草场管理中更应该侧重生态系统的弹性管理，而不仅仅是关注作为阈值的承载力本身。

在实现草原可持续发展方面，国内学者对于草原可持续发展能力的研究还处于起步阶段，该领域的概念、评价理论及其方法尚未达成共识。在理论研究方面，徐宜国（2009）将影响草原可持续发展的因素归纳为自然环境因素和社会环境因素，邓波（2004）认为草原可持续发展是生态、经济与社会可持续性三者的平衡，秦丹（2004）认为草原退化不但与各种不合理利用直接因素有关，而且与社会经济体制、人口压力、生物生态学等深层次因素相关。汪诗平（2006）讨论了天然草原持续利用在理论和实践方面存在的问题，如草原退化的面积和程度、造成退化的原因和权重等不确定性问题、载畜量及其估测的理论基础问题、"围封转移"和"春季休牧"措施的理论基础，以及草原生态系统多功能等问题。在评价方法上，相关学者（哈鸿儒，2006）通过实地调研和定性定量分析相结合的方法，研究草原可持续发展问题。刘兴元等（2009）基于草原生产力指数、草地生态服务价值指数、草地资源分类指数等对草地资源进行分类。颉茂华和塔娜（2013）运用"三重能力"的变权综合评价方法对我国的伊犁草原、锡林郭勒草原、那曲高寒草原及祁连山草原的可持续发展能力进行评价，认为只有在经济能力、资源生态能力及管理能力等三重能力互惠共生和协调发展的情况下，草原才具有较强的可持续发展能力。刘兴元等（2011）从草地生态系统的"三生功能"出发，综合考虑草地的地域和空间的异质性以及区域经济发展水平的差异性，选择水源涵养、侵蚀控制、废物处理、滞留沙尘、生物多样性保护、养分循环、固定CO_2、释放O_2、消减SO_2、畜牧生产、文化传承和休闲旅游12项贡献率较高的功能指标，建立了适宜于草地生态系统服务功能经济价值的评估体系和方法。

在协调多方利益方面，杨立华和欧阳赟（2012）通过对草原管理的智能体模拟分析发现，在特定条件下，策略人的策略决策深刻影响个体的具体行动，进而影响个体的行动结果，如草原的可持续发展；进而探讨了通过政策或制度安排来利用策略人的策略决策谋取预期政策或制度收益的可能性和其现实响应，并根据四种典型的策略决策（策略性投票、掩盖性策略决策、交易性策略决策和学习性策略决策），相应地讨论了策略约束性、保护性、诱导性及学习性四种政策或制度安排。同时，杨立华（2011）以草原治理为例，构建多元协作性治理的简单博弈模型，认为该治理模型在中国的实践应用有助于扩大参与主体，但也会因此导致多元主体之间的矛盾和冲突，增加多元协作性治理的困难。

从社区管理的角度，张倩（2013）提出，现有的自然资源管理多以行政单位定义社区，而忽略了社区与自然资源利用在不同尺度上的契合，导致政策实际的执行效果达不到预期目标，并以内蒙古荒漠草原的一个嘎查为例，结合奥斯特罗姆提出的公共池塘资源的多层次分析，认为基于社区的草原管理在操作选择过程中也需要分层次定制占用、提供、监督和实施的规则，通过建立一个基于社区的草原管理等级框架，保证草原管理与当地气候条件和畜牧业需求相契合，从而实现可持续的草原利用与保护。王晓毅（2009）认为，原有的“一刀切”的草原政策和村民排除在政策制定和实施以外的做法是导致草原退化的重要原因，也是草原恢复保护政策难以实现其预期目标的原因；以克什克腾旗皮房村民组民主协商进行草原管理为例提出，草原环境保护政策应以鼓励村民参与和发挥村民集体行动的能力为核心。

在草原管理资金投入方面，孟慧君等（2006）系统分析和探讨北方草原生态建设投资面临的问题，即草原生态经济投资不足、生态投资结构不佳、投资效益不高、牧民增收效果不明显、投资机制不完善、草原生态“边建设边破坏”、“三牧”政策尚需深入研究与贯彻；揭示其问题与困惑的实质，在于草原生态建设投资与牧区经济社会发展缺乏应有的互动，在于牧区工业化、城市化、草原畜牧业与草原生态建设的相互对立；进而提出，加强草原生态建设投资的优化管理，旨在摆脱困惑，实现草原牧区“生态改善、牧区繁荣、牧业增效、牧民增收”的多维目标的共荣，以推进草原牧区全面建设小康社会的进程。

在科技发展方面，侯向阳等（2004）运用层次分析法分析和评价了西部草业科技发展重点，认为目前西部草业科技发展的重点应为草地退化和调控机理、生态系统的界面与耦合、优质丰产抗逆牧草新品种选育、草地放牧管理等。

从完善草原法律体系的角度，冯学智和李晓棠（2013）对我国草原法律及地方法规进行了梳理，认为草原法律的完善需要与社会发展相一致，要在草原执法工作中弱化草原法律体系不完善而带来的偏差，同时要积极呼吁从中央、部门、地方多个层面尽快完善草原法律体系。万政钰和刘晓莉（2010）基于2002年年底修订后的《草原法》，对法律文本本身的价值、技术和实效进行了评价，认为存在包括草原立法的正义价值存在欠缺、逻辑结构不尽合理、刑事法律法规不够完整、草原犯罪的刑事责任难以实现或不能实现、草原违法的行政处罚数额幅度偏大等在内的几大问题。

2.4 草地区域管理实践研究

2.4.1 国外区域管理案例

1. 澳大利亚的牧场租赁计划

牧场租赁通常是由政府和租赁者（农民）签订合同，合同通常规定一个特定的时间（25～50年），指出必须用来放牧的特定地区——通常是用来放养畜牧群，但是昆士兰州为放牧而发布的租赁既要用来种植又要用来放牧。在南澳大利亚、西澳大利亚和北部地区，立法机关考虑到了其他一些用途，如园艺和农田度假等（表2-5）。

表2-5 各政府立法中牧场租赁的目的

地域	法案	立法目的
新南威尔士州	《1901年西部土地法案》	租赁必须用于牧养目的
昆士兰州	《1994年土地法》	租赁必须只能用于其公布的目的，牧养目的的期限租赁只用于农业放养目的
南澳大利亚	《1989年牧场土地管理和保护法案》	除非牧养委员会事先批准，否则除了牧养之外，租赁者没有权利使用土地做其他事情，牧养的目的意味着牧养储备和其他辅助的目的
西澳大利亚	《1997年土地管理法案》	除了牧养，牧场在没经过允许的情况下是不允许使用的。牧养目的意味着对批准的放牧进行商业的放养，农业、园艺或与批准的放养相关的其他的土地使用，以及辅助活动
北部地区	《1992年牧场土地法》	租赁必须用于牧养目的。牧养目的意味着持续的商业放养或农业或者其他对牧养企业至关重要的非主要用途，包括用于喂养储备的农产品和田园旅游，如农田度假
新西兰	《1948年皇家牧场土地法》	牧场使用期限的总体要求阐明租赁必须用于牧养目的

资料来源：宋洪远（2006）

司法部门对运营牧场企业的租赁者提出了具体的要求，包括担负一定的责任并遵循土地的管理。通常情况下，立法部门会阐明总体义务，然

后在政府代表和个体之间就特殊资产要求进行谈判。租赁者的运营要遵循以下一些要求：①在租赁与有效的土地管理实践一致的情况下运营企业；②防止土地恶化；③在财政资源允许的条件下，努力提高土地状况。另外，针对不同的管理系统建立了专门的租赁、租金和审查体系。

2. 亚洲开发银行资助项目的成功经验

基于已经实现成功开发的项目，亚洲开发银行对成功案例的相似特征进行了总结，这些项目的主要特征不一定是草地改良项目，但它们的开发计划应有双重目标，即增加收入和减少土地退化。其中个别经验对于其他项目的执行具有一定的借鉴意义，如下所示。

（1）这些项目有政府政策的指导，而这些政策都是在公共咨询之后制定出来的，反映了农村地区希望克服的问题和事项，从而改善当地人民的生活，保护当地的环境。

（2）这些项目都是建立在对自然资源能力的合理且实际的评估之上。

（3）这些项目都有着双重目标——增加收入和保护及恢复生态。

（4）开发活动由公众自己来确定、规划，若有需要，由当地部门提供技术支持。

（5）在设计、执行和监控过程中，农村社会的每个成员都有参与，包括妇女和儿童。

（6）各级政府及目标群体的所有成员了解和明确开发预算。

（7）开发资金或物资直接下达给目标群体，资金避开下级政府，直接存入当地的银行账户。

（8）项目执行在时间和安排上有一定的灵活性，由目标群体根据自己的资源情况，特别是劳动力的情况来安排。

（9）技术支持由当地相互合作的部门机关提供，如有需要，技术支持由当地受益者来组织和管理。

（10）技术管理培训课程不是一次性的，组织形式最好是以成人教育为基础，如农民田间学校等，课程重复的次数取决于季节和当地的劳动力文化水平。

（11）相关研究发现，开发对于女性劳动力需求的影响通常是负面的，

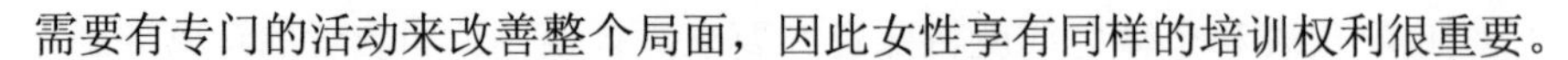

需要有专门的活动来改善整个局面，因此女性享有同样的培训权利很重要。

（12）从长远来看，以学龄儿童为中心的环境教育计划具有深远的意义，因为他们可能会成为当地环境的下一代管理者。

（13）公众运用简单的生活环境改善指数监控自身，这些指数可以融入地区及国家数据库，关键的成功指数在一开始就被设计出来，包含在项目中，所有的利益相关者都知道怎样测定和汇报短期及长期的成功。

（14）监控的关键是成果（主要是考察环境状况和贫困程度的改善），而非投入（宋洪远，2006）。

2.4.2 国内区域实践

杨邦杰等（2011）认为新疆草原自然保护区的建设和管理存在资金缺乏、周边社区发展与保护区建设的矛盾、资源类型与管理权限划分不清三大矛盾，认为应当通过加快推进草原管理体制改革，制订全国草原自然保护区发展建设规划，建立草原保护区长效投入机制，强化草原自然保护区科技支撑能力，以及加强草原监理体系建设等措施，推进草原保护区的建设和管理。

杨邦杰等（2010）通过对内蒙古自治区牛奶产品加工企业、畜牧业养殖基地、农牧业示范园区和特色产业加工基地的考察，以及对退牧还草项目区草原生态恢复情况的调查，提出加强草原保护建设力度，促进农牧民增收的政策建议；并依据草原保护和建设中存在的问题，提出具体的对策，包括将草原确定为我国国土主体和重要的战略资源，深化推进草原综合改革，推进草原承包经营，统筹农牧区协调发展，加强草原监理体系建设等。

达林太和娜仁高娃（2010）从内蒙古草原的过牧理论、过牧理论引发的退化理论，以及为其治理所实施的各类理论工程项目入手，揭示了造成草原生态环境恶化的人文因素，提出牧区建设和发展的整体性理论。他们认为，内蒙古草原牧区的建设发展需要充分考虑蒙古高原自然经济条件和在上千年历史过程中自发形成的蕴含着游牧文化的畜牧业布局，合理利用草地资源的游牧文化，这对于今天的管理依然有指导意义。

宋丽弘和唐孝辉（2012）对内蒙古草原的碳汇潜力和发展路径进行了分析，认为应当从国内与国外两个层面寻找路径：国内路径应本着完善地

方行政法规，促进草原增汇措施、碳汇评估及碳汇交易的法制化；国际路径以“清洁发展机制”为依托，构建草原碳汇机制，搭建国际合作平台，完善草原生态效益补偿制度。

蒲莉妮和郭宏远（2012）认为传统的生产方式导致甘南藏族自治州草原生态系统发生了逆向演变（植被严重退化、沙化加剧、盐碱化加重、生物多样性锐减等），现有的草原生产能力已经不能满足畜牧业发展的需求，有必要采取措施促进草原畜牧业生产方式向生态型、产业型转变。

王建兵等（2013）对甘肃中西部干旱、半干旱地区牧民对草原管理政策的认知程度进行了问卷调查。结果表明，牧民对草原政策的认知受到牧民文化程度、家庭收入水平、个人外出打工情况等多种因素的交互影响。因此，在加强干旱、半干旱地区牧区农业技术的推广、强化基础设施建设、明确草原经营权和使用权，以及加强草原保护宣传的过程中，应当注重牧民的这种认知特点，并据此设计相应的可被牧民认同的实施方案。

蔡虹等（2013）对甘肃省天祝藏族自治县抓喜秀龙乡进行了实地调研，运用参与式农村评估法，收集、整理了该区域的传统放牧方式、气候变化感知和本土应对策略，分析了该区域草地管理决策和科学研究的公众参与现状，结合调查结果提出建议，认为在青藏高原高寒草地可持续管理中应充分吸纳牧民的本土知识，增强牧业社会应对气候变化的能力。

冯宇等（2013）利用ETM影像数据和地理信息系统技术对呼伦贝尔草原生态功能区三期（2000～2003年、2004～2007年和2008～2011年）的防风固沙功能重要性进行了评价，并分析了各单因子重要性和综合重要性的动态变化及空间分布格局。结果表明，植被覆盖度是影响防风固沙功能重要性的关键因子。2000～2011年，呼伦贝尔草原生态功能区的防风固沙功能重要性呈先升高后降低的趋势；空间分布由东向西逐渐升高，沙带地区防风固沙功能重要性较高，温性草甸草原区和温性草原区防风固沙功能重要性较低。

曹叶军（2013）对锡林郭勒盟草原生态补偿的现状和补偿效能进行了研究，认为当前草原生态补偿存在补偿政策不稳定、补偿主体不全面、补偿标准过低、补偿内容单一和补偿时机不合理等问题，主张按照公平性、全面性、内生性和规律性的原则，从促进草原生态补偿多元化和制度化两方面完善锡林郭勒盟草原生态补偿，实现由外援型生态补偿向内生型生态补偿的转变，即由生态中心导向向协同发展导向的补偿模式的转变。

洪冬星（2012）对青海省刚察县、青海省星海县、新疆维吾尔自治区哈密市、新疆维吾尔自治区昌吉市、内蒙古自治区西乌珠穆沁旗及内蒙古自治区乌拉盖管理区6个在西部草原牧区具有典型代表特点的旗县（开发区）的农牧户开展入户调查研究，通过定性定量统计分析对调研区实施的草原建设措施现状和绩效进行了初步评价，并对可能影响农牧民参与草原生态补偿意愿的社会、经济、生理和认知因素进行了相关分析和Tobit模型回归分析，研究了牧户的认知变量和社会经济变量对其参与草原建设保护行动的影响。

2.5 启示：矛盾与问题

综上所述，对于中国西部草原的研究大概存在以下几方面的情况。

1. 草地资源建设与发展方面

对草原生态治理目前面临的形势缺乏正确和清醒的认识；畜牧业基础设施建设薄弱；对退牧还草支持明显不够；资金支持机制还没有切实建立；生态移民区配套设施建设比较薄弱；科技服务不足，发展集约养殖困难；防灾减灾机制不完善，治理效果得不到保护；科研培训经费严重短缺，保种资金缺乏，草原补播改良投资跟不上；牧草病虫防治经费缺乏，草原资源监测工作经费不足；饲草料不能均衡补给。

2. 生态补偿和移民方面

补偿标准低；采取按户补偿的方式容易产生不平衡；在外的游牧户没有纳入生态移民安置范围；后续产业短期内很难发展起来，移民户处于就业无门的状态；牧民融资困难，草场建设投入能力不足，迁移户发展资金严重不足；无草场、无牲畜、无文化牧户和移民迁居牧户的经济状况和生存状况很差，急需政府支持。

此外，缺乏对现有草原生态恢复和建设政策可能产生的负面影响的研究。现有政策措施和项目建设已经在很大程度上对草原资源和生态环境的恢复产生了积极的影响，然而一些草原改良项目的社会经济成本对某些牧

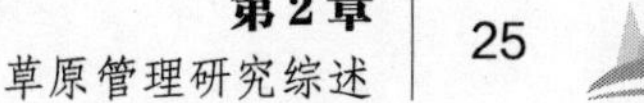

区的经济社会发展和牧民生活也有可能产生负面的影响，如果缺乏对这些负面影响的考察可能会削弱这些政策措施的预期效果。在讨论生态移民的问题时，缺乏对于牧民的心理承受能力、民族传统文化保护、定居模式的微观建构、城市化进程中民族问题及社区民族工作、定居政策及实施情况的研究。

3. 制度建设方面

地界纠纷不断；草原产权制度需要进一步完善；基本草原保护制度不完善；草原监理难度大；政策稳定性不够；执行机构能力较弱；草场保护力度不够；草场流转机制不完善。同时，缺乏对于其他社会保障制度的关联性研究，例如，在农牧民草原保护和恢复的支付意愿研究方面，由于农牧民在补偿政策及其风险的理解方面具有不确定性，或者存在信息不对称等问题，农牧民不愿意如实地给出期望的意愿支付价格（甚至给出零支付意愿）。

4. 国外草地保护的经验借鉴

应当采取具体的措施保障民众实现全过程参与以及公众监控进展；明确各级管理者、风险承担者的任务和责任；明确资金投入的去向和效果。

第 3 章

草原管理政策分析

3.1 国家层面草原管理政策法规综述

遏制草原生态恶化的趋势，是当前维护国家生态安全和实现经济社会可持续发展迫切需要解决的一个重要问题。相关研究指出，20 世纪 80 年代中期，国家对内蒙古、新疆、青海、西藏等北方草地的投资约为 47 亿元，平均每公顷草地投资为 15 元。而自 1949 年以来，若以 37 年平均分配，每年每公顷草地平均 0.3 ～ 0.45 元。到了 90 年代增加到典型草原每公顷年投资 0.75 元。20 世纪 90 年代后期，国家实施的西部大开发战略，推动了草原生态建设，也是草原建设投入最多的时期。2002 年 12 月 16 日正式批准在西部 11 个省份实施退牧还草政策，把草原生态保护建设提到重要的议事日程。

国家对草原生态建设的重视程度越来越高，在促进草原可持续发展方面提出了一系列的政策法规，下面主要从政策和法律法规两个方面综述我国 21 世纪以来主要的草原管理举措。

3.1.1 国家层面草原管理的主要政策

表 3-1 总结了 2005 ～ 2014 年中央一号文件中关于草原保护年度工作的部署，可以得到近十年我国草原政策的变化，表中政策内容重点的变迁

包括以下几个方面。

（1）草原政策的重心逐渐从保护到补助奖励转变，前期的政策重心在于草原的保护，强调生态效益，后期的政策逐渐开始兼顾牧民的利益，补偿手段也逐渐多样化。

（2）草原政策从单一的草地保护向多元、多层次的配套体系发展，围绕退牧还草，逐渐强调技术推广层面、制度层面和监管层面的政策。

（3）对于草原保护的重视程度逐渐加强，从政策实施的范围、力度和手段都反映了中央对于草原保护越来越重视。

表3-1　2005～2014年中央一号文件草原政策相关摘录

年份	草原保护政策
2005	进一步加强草原建设和保护，加快实施退牧还草工程，搞好牧放牧、轮牧休牧等生产区水利建设
2006	继续推进退牧还草、山区综合开发；建立和完善生态补偿机制
2007	完善森林生态效益补偿基金制度，探索建立草原生态补偿机制；加快实施退牧还草工程
2008	建立健全森林、草原和水土保持生态效益补偿制度，多渠道筹集补偿资金，增强生态功能；落实草畜平衡制度，推进退牧还草，发展牧区水利，兴建人工草场
2009	扩大退牧还草工程实施范围，加强人工饲草地和灌溉草场建设；提高中央财政森林生态效益补偿标准，启动草原、湿地、水土保持等生态效益补偿试点
2010	加大力度筹集森林、草原、水土保持等生态效益补偿资金；切实加强草原生态保护建设，加大退牧还草工程实施力度，延长实施年限，适当提高补贴标准；落实草畜平衡制度，继续推行禁牧休牧轮牧，发展舍饲圈养，搞好人工饲草地和牧区水利建设
2011	加强水利建设，坚持民生优先；抓好水土保持和水生态保护
2012	扩大退牧还草工程实施范围，支持草原围栏、饲草基地、牲畜棚圈建设和重度退化草原改良；加强牧区半牧区草原监理工作
2013	探索建立严格的工商企业租赁农户承包耕地（林地、草原）准入和监管制度；加快推进牧区草原承包工作，启动牧区草原承包经营权确权登记颁证试点；继续实施草原生态保护补助奖励政策
2014	继续在陡坡耕地、严重沙化耕地、重要水源地实施退耕还林还草；加大天然草原退牧还草工程实施力度，启动南方草地开发利用和草原自然保护区建设工程

3.1.2　国家层面草原管理的主要法律法规

从表3-2可以看出我国对于草原管理的法律法规处于不断的完善和细化过程中，自从《中华人民共和国草原法》实施以来，各种实施细则和意见不断出台，体现了国家对于草原管理法律法规建设的重视。

关于草原可持续发展的法律法规主要为《国务院关于加强草原保护与建设的若干意见》《国家草原生态保护补助奖励政策》《完善退牧还草政策》，其核心内容包括以下几个方面：

（1）建立基本草地保护制度；

（2）实行草畜平衡制度；

（3）推行划区轮牧、休牧和禁牧制度，实施禁牧补助、草畜平衡奖励及生产性补助。

表3-2　2002～2011年国家制定的有关草原保护的相关法律法规

年份	名称	主要内容
2002	《中华人民共和国草原法》（修订）	规定草原属于国家所有，由法律规定属于集体所有的除外。国家所有的草原，可以依法确定给全民所有制单位、集体经济组织等使用。草原承包经营权受法律保护，可以按照自愿、有偿的原则依法转让
2002	《国务院关于加强草原保护与建设的若干意见》	建立基本草地保护制度；实行草畜平衡制度；推行划区轮牧、休牧和禁牧制度
2003	《关于进一步做好退牧还草工程实施工作的通知》	建立基本草地保护制度；实行草畜平衡制度；推行划区轮牧、休牧和禁牧制度；完善实施方案，落实工程项目；推进家庭承包，明确权利义务；加强监督管理，实现草畜平衡；强化科技支撑，转变生产方式
2005	《中华人民共和国畜牧法》	国家支持畜牧业发展，发挥畜牧业在发展农业、农村经济和增加农民收入中的作用。国家帮助和扶持少数民族地区、贫困地区畜牧业的发展，保护和合理利用草原，改善畜牧业生产条件
2006	《草原征占用审核审批管理办法》	加强草原征占用的监督管理，规范草原征占用的审核审批，保护草原资源和环境，维护农牧民的合法权益
2010	《国家草原生态保护补助奖励政策》	制定了草原生态保护补助奖励的具体标准，具体包括禁牧补助、草畜平衡奖励及生产性补助
2011	《关于促进牧区又好又快发展的若干意见》	指出了促进牧区又好又快发展的意义、方针及总体要求，并给出了牧区新时期发展的意见，主要包括：加强草原生态保护建设，提高可持续发展能力；加快转变发展方式，积极发展现代草原畜牧业；促进牧区经济发展，拓宽牧民增收和就业渠道；大力发展公共事业，切实保障和改善民生；切实加强对牧区工作的组织领导
2011	《完善退牧还草政策》	给出了完善退耕还草政策的要求，包括：合理布局草原围栏；配套建设舍饲棚圈和人工饲草地；提高中央投资补助比例和标准；饲料粮补助改为草原生态保护补助奖励

3.2　黑河流域草原管理政策法规

由于不存在流域层面的行政单元，因此对流域草原管理政策的综述，以黑河流域行政单元（地级市）占主体的甘肃省、内蒙古自治区、青海省草原政策为例介绍。

3.2.1 甘肃省草原管理政策

1.《甘肃省草原条例》

《甘肃省草原条例》由甘肃省第十届人民代表大会常务委员会第二十六次会议于2006年12月1日通过，自2007年3月1日起施行。其主要内容包括：条例的实施总则，具体包含实施范围和执行部门等；草原规划建设部分，具体包含草原规划制定执行、草原调查、技术推广及灾害防治等内容；草原的所有制属性和承包经营管理；草原保护利用的详细说明；违反条例的法律责任认定。

该条例对草原管理的诸多方面都有明确的规定和说明，其中有关草原可持续发展的政策规定主要包括以下几个方面。

（1）草原使用者或承包经营者饲养的牲畜量不得超过核定的载畜量，保持可利用饲草饲料总量与其饲养牲畜所需饲草饲料量的动态平衡。

（2）草原使用者和承包经营者应当改变传统畜牧业生产方式，采取禁牧、轮牧、休牧和舍饲圈养等措施，提高草原的综合生产能力。

（3）禁止在草原上铲挖草皮、泥炭，防止造成新的植被破坏、草原沙化和水土流失。

（4）禁牧、休牧具体办法按国务院和省人民政府的规定执行。

（5）草原植被恢复费的具体标准由省人民政府依照国家有关规定制定。

2.《甘肃省草原禁牧办法》

《甘肃省草原禁牧办法》已经于2012年11月20日甘肃省人民政府第117次常务会议讨论通过，自2013年1月1日起施行，其主要内容：一是规定对严重退化、沙化、盐碱化、荒漠化的草原和生态脆弱区、重要水源涵养区的草原实施禁牧；二是规定由县级以上人民政府草原行政主管部门及其草原监督管理机构负责组织实施禁牧工作，所需工作经费列入本级财政预算；三是规定由省人民政府草原行政主管部门根据草原生态预警监测情况，划定草原禁牧区；四是将草原禁牧工作纳入各级人民政府目标管理责任制，层层签订责任书；五是规定乡镇人民政府在实施国家草原生态保护补助奖励机制政策的村设置草原管护员，并明确规定了草原管护员应当履行的职责；六是规定县级以上人民政府草原行政主管部门和乡镇人民政府健全草原禁牧区的巡查制度、举报制度、情况通报制度和禁牧补助资金的公示制度等。

3. 草原生态补助奖励政策

2011 年 7 月 31 日，甘肃省在其召开的全省落实草原生态保护补助奖励政策会议上，确定了草原生态补助的目标。可利用草原面积 2.41 亿亩[①]，中央财政下达草原生态保护补助奖励资金 11.4 亿元，主要包括禁牧补助、畜牧良种补贴、牧草良种补贴和牧民生产资料综合补贴等。确定禁牧草原面积为 1 亿亩，另外 1.41 亿亩草原在 3 年内完成 225.7 万个羊单位的减畜任务。

具体的措施包括以下几个方面。

（1）根据国家核定，甘肃省禁牧草原面积为 1 亿亩，其中青藏高原区 1727 万亩，黄土高原区 4156 万亩，西部荒漠区 4117 万亩，禁牧补助为 6 亿元。通过科学测算，三大区域的禁牧补助标准确定为青藏高原区每亩 20 元、黄土高原区每亩 2.95 元、西部荒漠区每亩 2.2 元，将通过“一册明一折统”直接发放到户。

（2）根据国家核定，甘肃省禁牧区域以外的 1.41 亿亩的可利用草原全部划定为草畜平衡区，草畜平衡奖励为 2.1 亿多元。据测定，这一区域超载牲畜 225 万个羊单位，平均超载率为 30.6%。为此，全省计划在 3 年内完成 225.7 万个羊单位的减畜任务。

（3）根据国家核定，甘肃省人工种草面积 2212 万亩，2011 年牧草良种补助为 2.2 亿多元，将用于更新现有留床牧草、扩大优良牧草种植、建设防灾保畜基地等。国家下达的畜牧良种补贴为 500 多万元，主要用于牦牛、山羊良种补贴，通过招标采购和养殖户选育调配，进行品种改良。

（4）根据国家核定，甘肃省以草原畜牧业为主要生产经营方式和收入来源的草原承包户为 22.1 万户，主要分布在 20 个牧业、半牧业县，实施牧民生产资料综合补贴，2011 年补贴资金 1.1 亿多元，按每户 500 元的标准，将补贴资金通过“一册明一折统”直接发放到户。

① 1 亩≈666.7m^2。

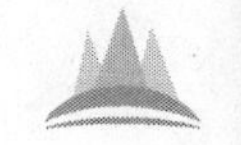

4.《甘肃省草畜平衡管理办法》

《甘肃省草畜平衡管理办法》已经于2012年9月18日甘肃省人民政府第114次常务会议讨论通过，自2012年11月1日起施行。关于草畜平衡管理的重点内容包括：各级人民政府应当普及草畜平衡知识，支持和引导农牧民采取人工种草、畜种改良、舍饲养殖等措施，减少草原载畜量，防止超载过牧；县级人民政府草原行政主管部门应当成立草原载畜量核定专家组，征求草原使用者和承包经营者的意见，核定草原载畜量，明确草原使用者或承包经营者的牲畜饲养量；按照国家草原生态保护补助奖励机制政策，对实现草畜平衡的农牧民给予奖励。奖励资金应当按照已承包到户（含联户）并实施草畜平衡的草原面积直接发放到户。

3.2.2　内蒙古自治区草原管理政策

1.《内蒙古自治区森林草原防火条例》

2004年3月26日，内蒙古自治区第十届人民代表大会常务委员会第八次会议通过《内蒙古自治区森林草原防火条例》，自2004年4月15日起施行。该条例强调预防和扑救森林草原火灾，保护森林草原资源，是草原管理的主要工作，并将草原管理的主要工作总结如下。

1）森林草原防火组织

组织有关部门和当地驻军、武警部队设立森林草原防火指挥部，负责本行政区域内的森林草原防火工作。受自治区人民政府或者所在旗县级人民政府委托，内蒙古大兴安岭林管局及其所属的国有林业局设立防火指挥部，负责施业区内的防火工作。旗县级以上人民政府森林草原防火指挥部下设办公室，负责处理指挥部的日常工作。

2）森林草原火灾的预防

各级人民政府负责组织划定森林草原防火责任区，确定森林草原防火责任单位，建立健全森林草原防火责任制度并定期进行检查。落实防火责任制应当贯彻属地管理为主的原则，任何单位和个人在落实防火责任方面，必须接受当地森林草原防火指挥部的监督。

3）森林草原火灾的扑救

任何单位和个人发现森林草原火情，必须立即进行扑救，并及时向当地人民政府或者森林草原防火指挥部报告。任何单位和个人应当无偿为报警提供便利。禁止谎报森林草原火情。

4）善后工作

森林草原火灾扑灭后，当地人民政府应当组织有关部门及时制订植被恢复计划，并组织实施；做好过火地的病虫害防治和火灾地区人畜疫病的防治、检疫，防止疫病的发生和传播。

2.《内蒙古自治区草原管理条例》

2004年11月26日，内蒙古自治区第十届人民代表大会常务委员会第十二次会议修订通过《内蒙古自治区草原管理条例》，自2005年1月1日起施行。其主要内容包括：草原管理条例实施的总则、草原权属及流转规定、草原规划制度、草原建设利用保护及监督及其应当承担的法律责任等。

该条例对草原管理的诸多方面都有明确的规定和说明，其中有关草原可持续发展的政策规定主要有以下几个方面。

（1）禁止开垦草原。对水土流失严重、有沙化趋势、需要改善生态环境的已垦草原，应当有计划、有步骤地退耕还草；已造成退化、沙化、盐碱化、荒漠化的，应当限期治理。

（2）对草原实行以草定畜、草畜平衡制度。草畜平衡核定由旗县级人民政府草原行政主管部门每三年进行一次，并落实到草原所有者和使用者。

（3）已经承包经营的国有草原和集体所有草原，依据核定的载畜量，由拥有草原使用权或所有权的单位与草原承包经营者签订草畜平衡责任书。未承包经营的国有草原，由草原使用者与旗县级以上人民政府签订草畜平衡责任书。未承包经营的集体所有草原，由草原所有者与苏木乡级人民政府签订草畜平衡责任书。

（4）自治区依法实行退耕、退牧还草和禁牧、休牧制度。禁牧、休牧的地区和期限由旗县级人民政府确定并予以公告。不得在禁牧、休牧的草原上放牧。

（5）实施草原建设项目以及草原承包经营者建设小面积人工草地需要

改变草原原生植被的，应当符合草原保护、建设、利用规划。旗县级以上人民政府草原行政主管部门应当加强监督检查。

（6）不得在下列草原地区建设旱作人工草地：年平均降水量在250 mm以下的；坡度20° 以上的；土质、土壤条件不适宜种植的。

（7）禁止在荒漠、半荒漠和严重退化、沙化、盐碱化、荒漠化、水土流失的草原及生态脆弱区的草原上采挖植物和从事破坏草原植被的其他活动。

3.《内蒙古自治区草畜平衡暂行规定》

为了加强草原的保护、建设和合理利用，促进畜牧业可持续发展，内蒙古自治区于2000年8月1日起实施《内蒙古自治区草畜平衡暂行规定》。规定强调实行草畜平衡制度应当贯彻增草增畜、提高质量、提高效益的畜牧业发展战略，坚持畜牧业发展与保护草原生态并重的原则。

1）草畜平衡核定

旗县畜牧行政主管部门应当每两年进行一次草畜平衡核定，即对饲草饲料总贮量进行测算，并确定适宜载畜量。草畜平衡核定以草原承包经营者或草原使用权单位为单位进行。未承包的机动草原的草畜平衡核定以草原所有者或草原使用权单位为单位进行。

2）草畜平衡管理

草原所有者或草原使用权单位必须与草原承包经营者签订草畜平衡责任书。牲畜饲养量超过适宜载畜量的，草原承包经营者必须采取以下措施：种植和贮备饲草饲料，增加饲草饲料供应量；采取舍饲，进行阶段性休牧或划区休牧轮牧；优化畜群结构，提高出栏率。

3）奖励与处罚

对积极进行草原建设，实现草畜平衡的草原承包经营者、草原所有者和草原使用单位，各级人民政府应当给予表彰和奖励。对于违反本规定，不签订草畜平衡责任书的，超载牲畜，抢牧、滥牧严重破坏草原的行为，旗县以上畜牧行政主管部门应给予警告和处罚。

4.《内蒙古自治区退耕还林管理办法》

2007年9月27日，内蒙古自治区人民政府第八次常务会议通过《内蒙古自治区退耕还林管理办法》，并决定于2007年12月1日起实施。本办法对包括退耕地还林、配套的荒山荒地造林和封山（沙）育林在内的退耕还林活动进行规范，其中有关草原可持续发展的政策规定主要有以下几个方面。

（1）旗县级以上人民政府应当采取措施，将退耕还林与基本农田建设、农村能源建设、生态移民、后续产业发展结合起来，巩固退耕还林成果。

（2）旗县级以上人民政府应当支持退耕还林应用技术的研究和推广，提高退耕还林科学技术水平。

（3）退耕地还林实行区域化规模治理，除整体移民以外，可以合理调整耕地，应当保留人均3亩以上的口粮田。

（4）旗县级人民政府财政部门发放粮食补助资金应当与退耕还林成活率、保存率挂钩。

（5）各级人民政府应当征求退耕还林者意见，直接发给退耕还林者和承担配套荒山荒地造林的单位，或者进行集中采购退耕还林种苗。

（6）政府财政部门每年应当安排专项资金作为退耕还林的地方配套资金，用于退耕还林的前期费、管理费和验收费。

（7）禁止在退耕还林实施范围内复耕和从事滥采、乱挖、放牧等破坏地表植被的行为。

3.2.3 青海省草原管理政策

1.《青海省实施〈中华人民共和国草原法〉办法》

2007年9月28日青海省第十届人民代表大会常务委员会第三十二次会议通过《青海省实施〈中华人民共和国草原法〉办法》，对本省行政区域从事草原规划、保护、建设、利用和管理等活动进行规范，具体包括如下几个方面。

1）草原权属

本省行政区域内依法确定给全民所有制单位和集体经济组织使用的国家所有的草原，由县级以上人民政府登记，核发使用权证，确认草原使用权。依法改变草原权属的，应当办理草原权属变更登记手续。此外，草原

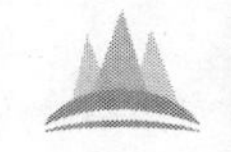

承包经营期限一般为五十年，期间草原承包经营者依照国家和省有关规定实施退牧还草后的草原承包经营权不变。

2）草原规划与建设

各级草原行政主管部门应依据上一级草原保护、建设、利用规划，编制本行政区域的草原保护、建设、利用规划。草原保护、建设、利用规划确需调整或者修改的，须经原批准机关批准。此外，县级以上人民政府草原行政主管部门应当会同相关部门，对草原的面积、等级、植被构成、生产能力、生态状况、自然灾害、生物灾害等草原基本状况实行动态监测且共享监测信息成果，并及时提供动态监测和预警信息。

3）草原利用

实行草畜平衡制度，逐步实施草畜平衡的各项措施。各级草原行政主管部门应当根据具体载畜量标准和本行政区域内草原基本状况、草地生产能力、动态监测结果，核定、公布草原载畜量，确定草原承包经营者或草原使用者的牲畜饲养量。草原承包经营者或草原使用者应当合理利用草原，采取人工补饲、舍饲圈养、加快牲畜出栏、优化畜群结构等措施，保持草畜平衡。

4）草原保护

实行基本草原保护制度，对基本草原实施严格保护管理。任何单位和个人不得破坏草原保护标志和围栏等草原保护设施，不得擅自占用或者改变用途。各级政府依法实行退耕、退牧还草和禁牧、休牧制度，不仅应加强疏林草地和灌丛草地的植被保护，禁止开垦草原，适时合理利用草地，还应强化对草原珍稀濒危野生动植物的保护。此外，各级人民政府应当建立和完善草原防火责任制，并将草原防火期定位为每年9月15日至翌年6月15日。

5）监督管理

县级以上人民政府草原行政主管部门草原监督管理机构的职责是：宣传贯彻草原法律、法规和规章，监督检查草原法律、法规和规章的贯彻执行；依法对违反草原法律、法规和规章的行为进行查处；负责办理草原承包经营权、草原使用权的登记、造册和发证的具体工作；负责草原权属争议的调解；负责草原动态监测工作，指导监督草畜平衡工作；协助有关部门做好草原防火的具体工作；办理其他有关草原监督管理的事宜。

2.《青海省草原生态保护补助奖励机制实施意见（试行）》

为加强草原生态保护，促进畜牧业经济和社会可持续发展，结合青海省草原牧区实际，青海省人民政府于 2011 年 9 月 28 日印发《青海省草原生态保护补助奖励机制实施意见（试行）》，并提出加快建立草原生态保护补助的奖励机制。

（1）实施天然草原禁牧补助政策，制定禁牧补助标准。以全省年平均饲养一个羊单位所需 26.73 亩天然草原作为一个“标准亩”，结合国家禁牧补助测算标准，确定各州禁牧补助标准和资金总额。各州结合实际，确定本地禁牧补助标准并落实到户。

（2）建立草畜平衡奖励政策。确定超载牲畜数量，制订减畜计划，核减超载牲畜，建立以草定畜、草畜平衡制度，实施草畜平衡管理。各级财政、农牧部门对履行减畜计划的牧民，按照每年每亩 1.5 元的测算标准给予草畜平衡奖励。

（3）落实对牧民的生产性补贴政策。根据畜牧业生产实际需要，落实对牧业生产所需的牦牛种公牛、绵（山）羊种公羊良种补贴，补助标准为：牦牛 2000 元 / 头，藏绵羊 800 元 / 只，绒山羊 800 元 / 只。此外，还要落实对牧区、半农半牧区人工种草所需籽种的补贴。

（4）建立绩效考核和奖励制度。实行目标、任务、资金、责任“四到州（地、市）”，对各地补助奖励政策落实、草原生态保护效果、补助奖励资金使用管理及配套政策制定落实等情况进行绩效考评。对工作突出、成绩显著的地区给予奖励，对完不成目标任务的予以通报批评。

第 4 章 黑河流域主要生境要素的空间分析

黑河流域是我国第二大内陆河流域，具有干旱区内陆河流域的一切特点。黑河位于祁连山和河西走廊的中段，介于 37º45′～ 42º40′N、96º42′～ 102º4′E，东西和南北跨度大，南北大约跨 5º，东西跨 6º 之多。黑河发源于南部祁连山区，干流分东西两大支流，东支俄博河（又称八宝河），自东向西流长 80 km，西支野牛沟，由西向东流长 190 km，东西两支汇于黄藏寺折向北流称为甘州河，流程 90 km 至莺落峡进入走廊平原，始称黑河，莺落峡以上为上游。黑河从莺落峡进入河西走廊，穿越正义峡（北山），进入阿拉善平原。莺落峡至正义峡流程 185 km，为黑河（干流）的中游。正义峡以下为黑河的下游，黑河流经正义峡谷后，北流 150 km 至内蒙古自治区额济纳旗境内的狼心山西麓，又分为东西两河，东河（达西敖包河）呈扇形注入东居延海（索果淖尔）；西河（穆林河）注入西居延海（嘎顺淖尔）。

黑河流域地貌类型多样，在大地构造上大体可分为南部祁连山地槽褶皱带、北部阿拉善台隆和北山断块带，以及河西走廊拗陷盆地三种基本单元，自北向南，海拔逐渐升高（图 4-1）。黑河流域位于欧亚大陆中心部位，远离海洋，周围高山环绕。气候干燥，降水稀少而集中，多大风，日照充足，太阳辐射强烈，昼夜温差大。由于受大陆性气候和青藏高原的祁连山－青海湖气候区影响，中下游的走廊平原及阿拉善高原属中温带甘－蒙气候区，该区根据干燥程度，可进一步分为中游河西走廊温带干旱亚区及下游阿拉善荒漠干旱亚区和额济纳荒漠极端干旱亚区，南部山区属于山地半湿润气候亚区。由于气候条件的不同，流域自然环境条件差异显著，流域内自东南向西北形成了链状山地—斑块绿洲—广域荒漠的生态系统。由此导致植被、土壤的地带性和非地带性的分布并存，也为草地类型的多样性奠定了环境基础。

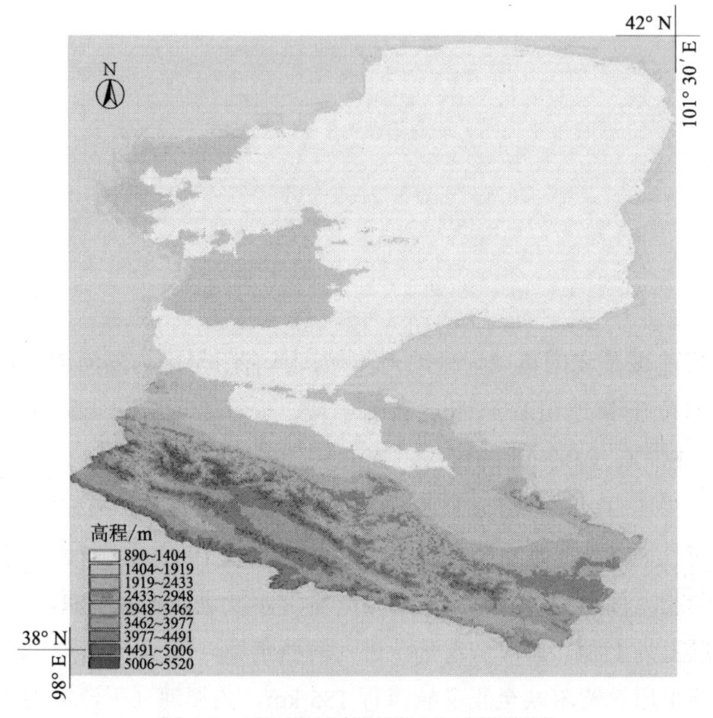

图 4-1 黑河流域高程图（文后附彩图）

流域土壤和植被具有显著的空间差异。可划分为 21 种土类，其中地带性土类 16 种，非地带性土类 5 种。黑河上游包括高山寒冷荒漠土壤系列、高山草甸土壤系列、山地草甸草原土壤系列、山地草原土壤系列和山地森林土壤系列，主要土类有寒漠土、高山草甸土（寒冻毡土）、高山灌丛草甸土(泥炭土型寒冻毡土)、高山草原土(寒冻钙土)、亚高山草甸土(寒毡土)、亚高山草原土（寒钙土)、灰褐土、山地黑钙土、山地栗钙土、山地灰钙土等。流域中游地区属灰棕荒漠土与灰漠土分布区。除这些地带性土类外，还有灌淤土（绿洲灌溉耕作土)、盐土、潮土（草甸土)、潜育土（沼泽土）和风沙土等非地带性土壤。在下游额济纳旗境内，以灰棕漠土为主要地带性土壤，受水盐运移条件和气候及植被影响，非地带性分布硫酸盐盐化潮土、林灌草甸土及盐化林灌草甸土、碱土、草甸盐土、风沙土及龟裂土等。

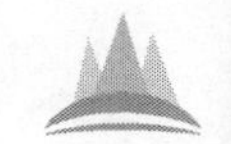

植被在水平分布上自南向北，大体为草甸草原、森林草原、荒漠化草原、草原化荒漠及荒漠植被带。垂直带谱极其分明，海拔4000～4500 m为高山垫状植被带，3800～4000 m为高山草甸植被带，3200～3800 m为高山灌丛草甸带，2800～3200 m为山地森林草原带，2300～2800 m为山地干草原带，2000～2300 m为草原化荒漠带。流域中、下游地带性植被为温带小灌木、半灌木荒漠植被。受河流水源和人类活动影响，中游山前冲积扇下部和河流冲积平原上分布有灌溉绿洲栽培农作物和林木，呈现以人工植被为主的景观。而在河流下游两岸、三角洲上与冲积扇缘的湖盆洼地一带，呈现荒漠天然绿洲景观。

地域经济类型呈现出山地牧业区—走廊绿洲农业区—高原牧业区组合特征，具有典型的农牧业分异和农牧交错特色。本书通过收集黑河流域的观测与调查数据，分析流域的自然特点，为理解草地的分布特征提供重要的前提条件。生境要素包括很多，在黑河流域影响植被分布和生产力的关键因素是气温和降水。由于空间和地形的限制，观测站点十分稀疏。利用有限的站点数据模拟草地的空间分布是不可能的，因此气温和降水的空间化显得十分必要。利用已有的站点，通过统计方法建立气温和降水与地理位置和海拔的关系，结合ARC/INFO空间分析模块，实现气温和降水的空间分布，为草地的潜在分布模拟奠定基础。

4.1 数据的获取与方法

收集研究区及其周边30个雨量站和分布在研究区的22个站点的气象数据，数据的观测时段为55年（1960～2014年），同时收集各站点的地理位置和海拔数据。在数字黑河网站下载分辨率为90 m的数字高程模型（Digital Elevation Model，DEM）。

4.1.1 黑河上游气温与降水的空间分布模拟

根据站点的分布，将流域划分为上、中、下游进行气温和降水的空间分布分析。黑河上游为海拔较高的山地占据，气温随着海拔的升高而降低，由于高海拔地形截留空中水汽，随着海拔的升高，降水量在增加。又因流域空间位置受海洋性气团的影响不同，降水量在东西部差别大，因此分布式的降水和气温模式的自变量选择海拔、经度和纬度。采用 SPSS 软件对黑河上游各点多年平均气温和降水数据与经度、纬度、海拔进行相关分析，并建立它们之间的关系式如式（4-1）和式（4-2）的形式。

$$P = a + bH + cY + dX \tag{4-1}$$

$$T = a + bH + cY + dX \tag{4-2}$$

式（4-1）和式（4-2）中，X、Y 分别为观测站点的经度和纬度；H 为观测站点的海拔；a，b，c，d 为回归系数。

4.1.2 黑河中游气温与降水的空间分布模拟

黑河中游地势平坦，分布在中游区域的观测站点相对较多，利用统计分析和插值分析两种方法对中游的气温和降水的空间分布进行模拟，结果比较发现，插值方法获取的精度要高，因此利用空间插值对中游的气温和降水进行空间化。插值方法使用普通克里金插值方法。

4.1.3 黑河下游气温与降水的空间分布模拟

黑河下游面积占整个流域的近 70%，分布在流域的站点只有 6 个。站点稀少，不具有统计意义，也无法进行空间插值，对于该区域的气温和降水空间分析依赖于站点的空间分布。根据下游的 6 个站点的空间分布，分析气温和降水的南北差异。

4.1.4 黑河流域潜在草地模拟

1. 黑河上游潜在草地模拟

利用降水量 200 mm 为阈值进行划分，分为两个区域：降水量小于 200 mm 的区域定义为荒漠化草原或荒漠；降水量大于 200 mm 的区域主要分布在上游。利用 GIS 空间分析软件，将上游草地现状分布从整个流域中提取出来，在降水量和气温空间化的基础上，用草地分布图层提取分布

区多年平均气温与降水的分布空间，根据水热条件的分布空间，界定边界函数（图 4-2），其上边界和下边界函数分别为

$$P_u = -1.464T^2 - 26.12T + 414.4 \tag{4-3}$$

$$P_l = 0.969T^2 - 11.23T + 294.0 \tag{4-4}$$

在 ARC/INFO 中利用空间运算，获取上游草地的潜在空间分布。

2. 黑河下游潜在草地模拟

下游为极端干旱的荒漠分布区，但是受地下水分布的影响，在河岸带有非地带性植被分布，将河岸带受地下水影响的区域做缓冲带，在缓冲带调查草地盖度和地下水。利用调查数据建立地下水与关键物种分布盖度的关系，其关系如下所示。

$$C = \frac{1}{\sqrt{2\pi}\sigma x} \exp\left[-\frac{(\ln x - \mu)^2}{2\sigma^2}\right] \tag{4-5}$$

式中，C 为草地盖度；x 为地下水深；μ 和 σ 分别为模型参数。利用地下水模型对地下水的空间分布进行模拟，获取黑河下游地下水空间分布，为下游潜在草地的模拟做数据准备。

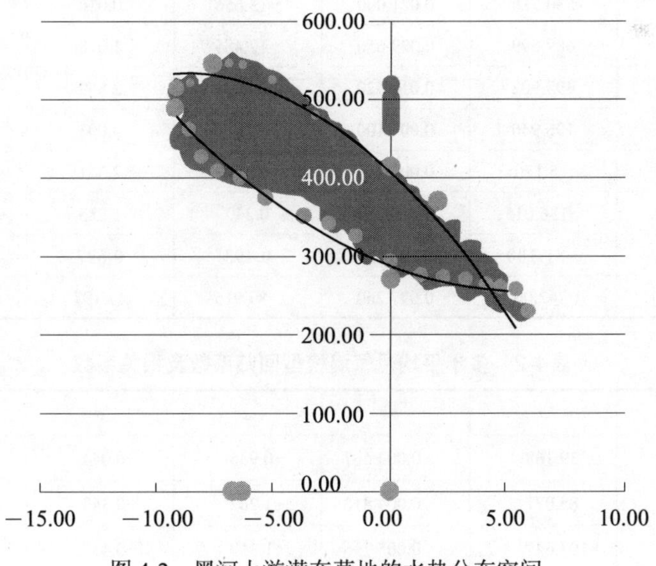

图 4-2　黑河上游潜在草地的水热分布空间

4.2 结果

4.2.1 上游分布式气温和降水模式及空间分布

分布式的降水和气温模式见表 4-1 和表 4-2。由表 4-1 可以看出，降水模型精度最佳的月份是 5 ～ 10 月，其多年平均 R^2 值都在 0.8 以上。在黑河流域上游，降水集中在 5 ～ 9 月，期间的降水量占全年降水量的 80%。所以线性回归方程被认为是合理的选择。由表 4-2 可以看出，气温模型精度各月的 R^2 值都在 0.8 以上。利用回归模型结合 GIS 空间分析方法，获得黑河上游各月和多年平均降水量和气温的空间分布，年平均降水量、年平均气温、最高月和最低月气温空间分布见图 4-3 和图 4-4。

表4-1 多年平均月降水模型回归系数与相关系数

时间	a	b	c	d	R^2
1 月	−19.811	0.000 260	−0.051	0.231	0.207
2 月	−70.701	0.001 103	0.221	0.626	0.331
3 月	−249.545	0.003 390	0.433	2.336	0.406
4 月	−16.862	0.004 009	−4.289	1.879	0.584
5 月	408.331	0.009 569	−12.540	0.869	0.810
6 月	530.716	0.021 000	−13.656	0.016	0.863
7 月	689.699	0.029 650	−12.485	1.018	0.870
8 月	495.902	0.018 520	−19.839	2.869	0.879
9 月	196.940	0.009 100	−15.049	4.003	0.856
10 月	−5.170	0.002 153	−5.737	2.341	0.841
11 月	−136.015	0.000 984	0.240	1.283	0.455
12 月	−81.180	0.000 480	0.493	0.627	0.166
多年平均	1 742.001	0.097 260	−87.915	17.197	0.861

表4-2 多年平均月气温模型回归系数及相关系数

时间	a	b	c	d	R^2
1 月	39.168	−0.004 467	−0.935	−0.044	0.893
2 月	85.077	−0.004 816	−1.262	−0.343	0.961
3 月	107.647	−0.005 736	−1.319	−0.457	0.981

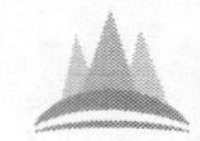

续表

时间	*a*	*b*	*c*	*d*	R^2
4月	68.609	−0.006 203	−0.248	0.642	0.986
5月	19.472	−0.006 538	0.131	0.008	0.983
6月	−58.924	−0.006 049	1.230	0.392	0.985
7月	−53.779	−0.005 750	1.382	0.294	0.988
8月	−28.831	−0.005 816	0.965	0.199	0.985
9月	−30.533	−0.005 264	0.596	0.296	0.981
10月	−2.415	−0.005 116	−0.101	0.213	0.972
11月	64.51	−0.005 668	−1.033	−0.165	0.958
12月	41.481	−0.004 511	−0.982	−0.039	0.887
多年平均	21.784	−0.005 49	−0.01	0.131	0.981

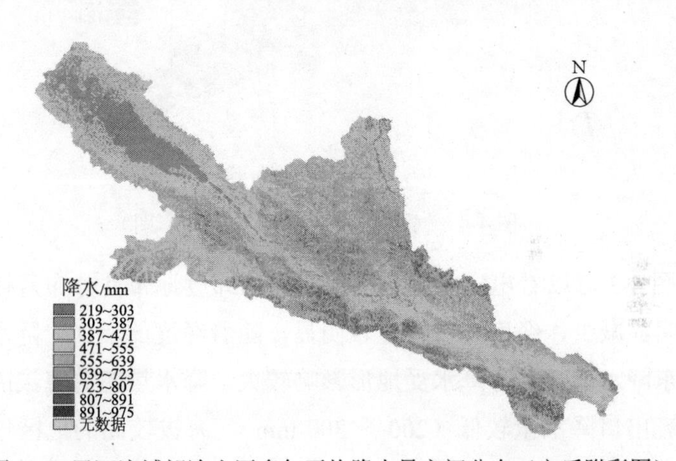

图 4-3　黑河流域祁连山区多年平均降水量空间分布（文后附彩图）

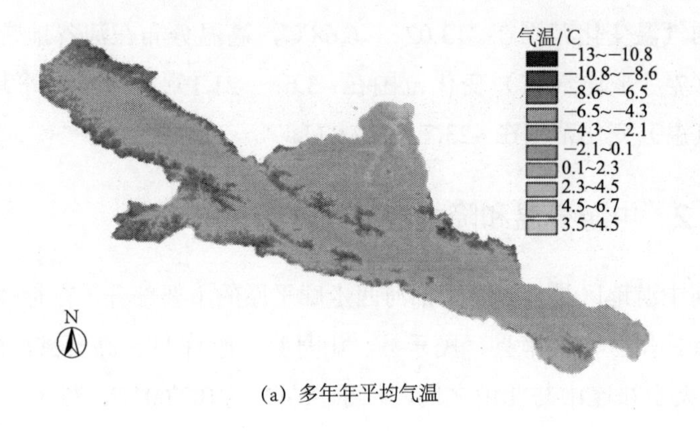

(a) 多年年平均气温

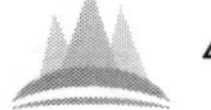

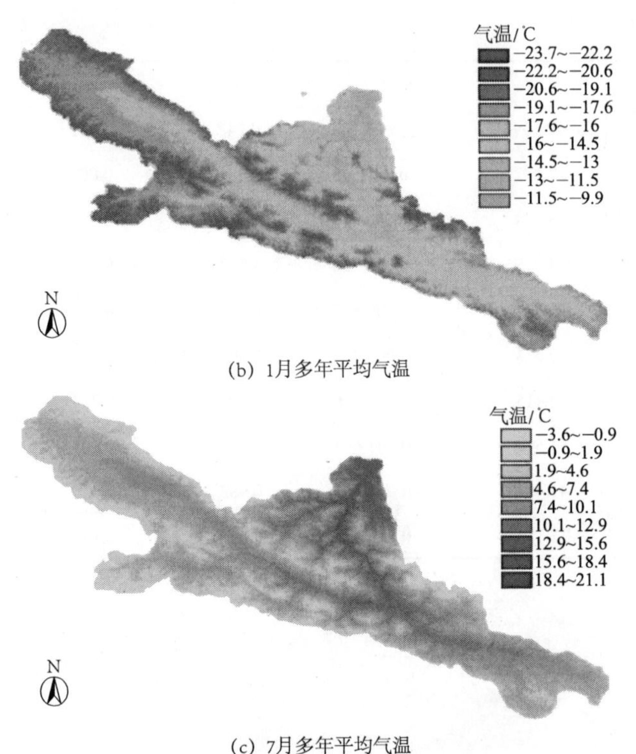

(b) 1月多年平均气温

(c) 7月多年平均气温

图 4-4　气温的空间分布（文后附彩图）

由图 4-3 可以看出：①降水呈现明显的纬度地带性分布规律，自南向北降水明显减少，经度地带性也较明显，随着经度的增加，降水增加，即由西向东降水增加；②降水受地形影响较大，降水量随着海拔的增加而增加，河流出口降水量较低（200 ～ 300 mm），海拔较高的地区年降水量较大（400 ～ 700 mm），个别地区可达到 700 mm 以上。由图 4-4 可以看出：①年平均气温变化范围在 –13.02 ～ 6.68℃，高温分布在河谷地带；②最高月气温（7 月平均气温）变化范围在 –3.6 ～ 21.1℃，最低月平均气温（1 月平均气温）变化范围在 –23.7 ～ –9.9℃。

4.2.2　中游气温和降水的空间分布

黑河中游地区是甘肃省中部河西走廊平原的主要部分，在行政区域上包括张掖市甘州区、肃南县、民乐县、山丹县、临泽县、高台县，面积约 1.7 万 km^2，夹于祁连山与北山之间，位于 96°42′～ 102°00′E、37°41′～ 42°42′N

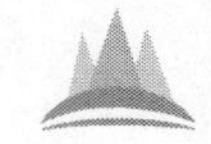

（图4-5），海拔在1300～2500 m。处在低山、荒漠、绿洲与沙漠的交汇处。

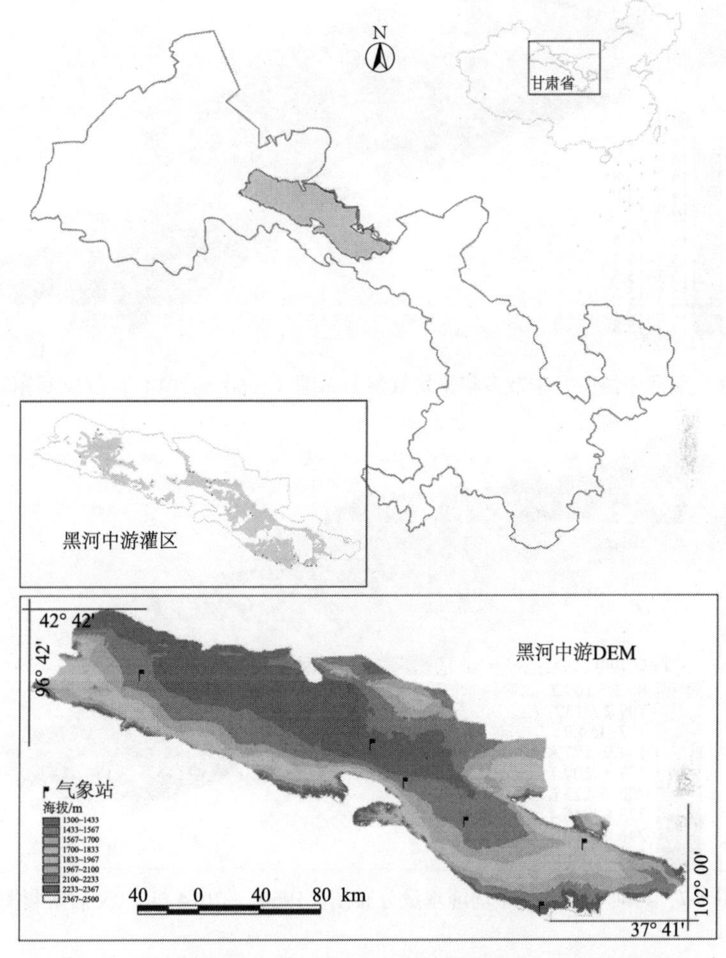

图4-5　黑河中游相对地理位置与数字高程模型（文后附彩图）

利用克里金空间插值方法获取中游气温和降水量的空间分布（图4-6、图4-7）。由图4-6可以看出：温度由东南向西北逐渐增加，多年平均气温在该区分布的范围为14.5～15.8℃，温度较高。降水与气温有着相反的趋势，即由东南向西北逐渐减少，多年平均降水由东南的292.2 mm减少到西北的86.3 mm，降水梯度较大（图4-7）。水热的组合使得该区域的地带性植被为中温带荒漠化草原植被，主要由超旱生灌木、半灌木或盐生、旱生肉质半灌木组成。野外物种调查显示，主要植物有白刺、红砂、珍珠猪毛菜、麻黄、白茎盐生草等。中游植被以人工植被为主体，大部分人工林为灌溉植被。

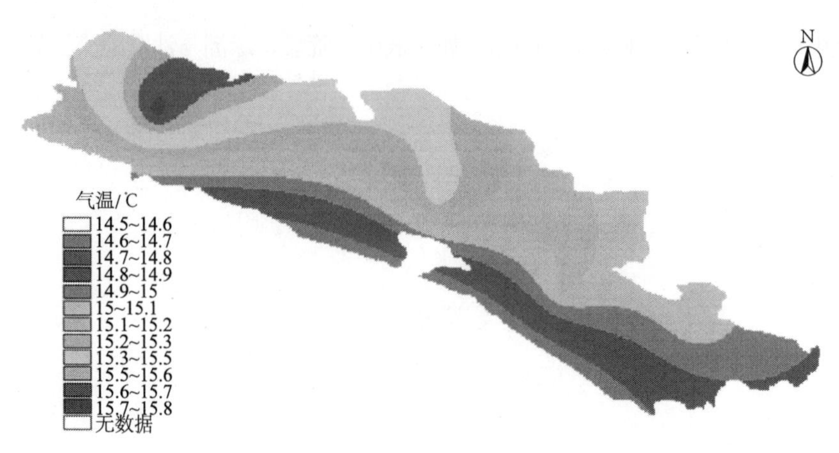

图 4-6　黑河中游生长季节多年平均气温分布图（1960 ～ 2014 年）（文后附彩图）

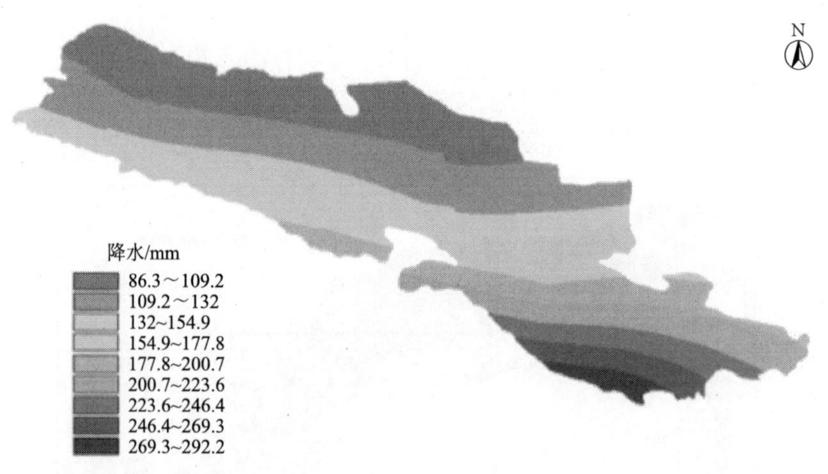

图 4-7　黑河中游多年平均降水量分布图（1960 ～ 2014 年）（文后附彩图）

4.2.3　下游气温和降水的空间分布特征

根据站点观测数据分析可知，除西部海拔较高的马鬃山外，在区域中的其他地区多年平均气温差别不大，6 个站点多年平均温度分布的范围为 4.4 ～ 9.3℃（图 4-8），最突出的特点是年较差大（图 4-9）；而降水量少且集中于 7、8 和 9 月份降落（图 4-10），多年平均降水量不超过 100 mm，随着纬度的增加，降水量逐渐减少（除马鬃山外），北部的额济纳旗降水量最少，多年平均降水量不超过 50 mm（图 4-11）。

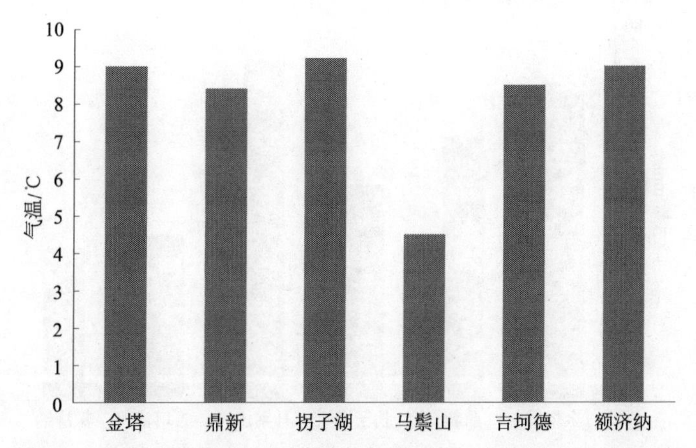

图 4-8　黑河下游 6 站点沿纬度多年平均气温变化图（由南到北）

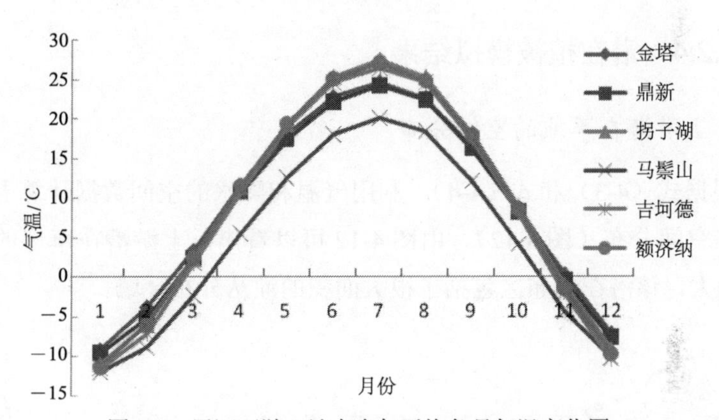

图 4-9　黑河下游 6 站点多年平均各月气温变化图

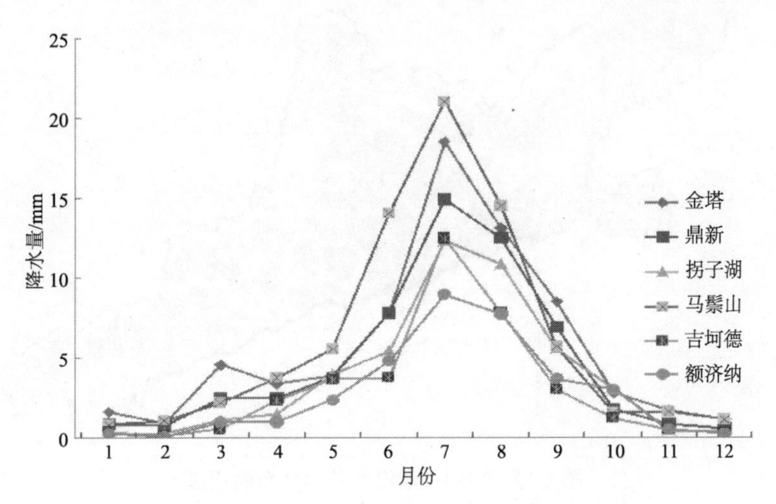

图 4-10　黑河下游 6 站点多年平均降水量各月变化图

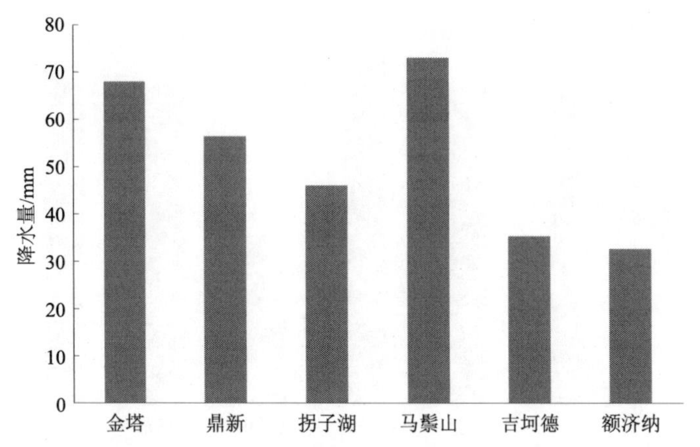

图 4-11　黑河下游 6 站点沿纬度多年平均降水量变化图（由南到北）

4.2.4　潜在植被模拟结果

1. 上游潜在草地的空间分布

根据式（4-3）和式（4-4），利用气温和降水的空间数据计算上游草地的潜在空间分布（图 4-12）。由图 4-12 可以看出，上游潜在草地的空间分布面积大，该潜在分布区包括了很大面积的灌丛分布区域。

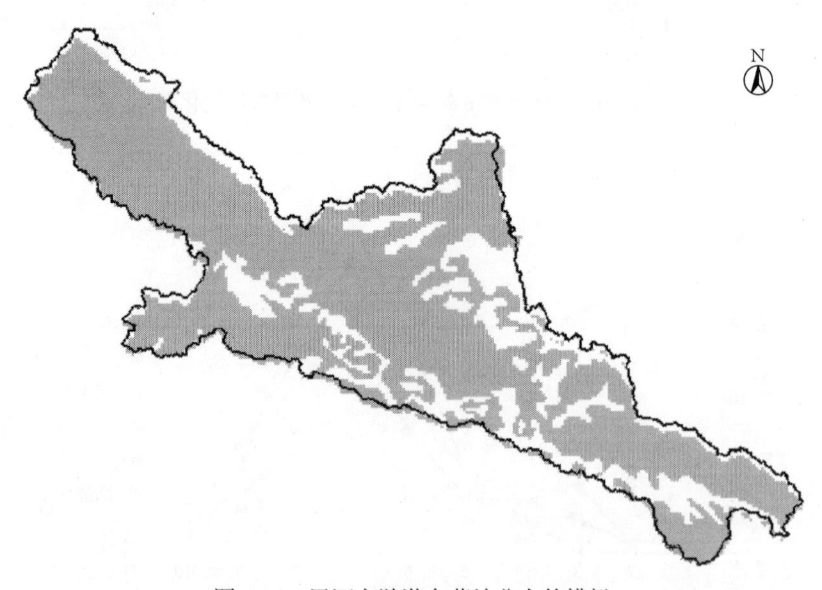

图 4-12　黑河上游潜在草地分布的模拟

2. 下游潜在草地的空间分布

利用地下水与植被盖度的关系，见式（4-5），结合地下水的空间分布对草地植被的潜在分布进行模拟，其空间分布见图4-13。柽柳作为大型牲畜的主要可食物种，其潜在分布的面积要远远超出现实的分布区域，该物种的减少主要是由于地下水下降，大面积的柽柳死亡。

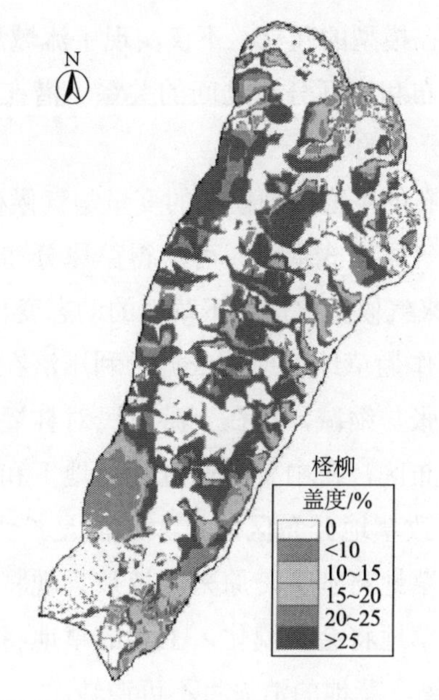

图4-13　黑河下游河岸缓冲区潜在植被分布（文后附彩图）

4.3　结论

（1）气温与降水是黑河流域的关键生境要素，对二者的空间化是草地植被潜在模拟的前提条件，空间化深受观测站点的限制。对于上游采用了统计方法，因为地形对气温和降水的再分配有控制作用。对于中游平坦区域，采用了克里金空间插值方法。对于下游，利用区域的站点的空间分布特征进行分析。

（2）不同的水热组合条件控制着植被的分布和生产力。上游地区高湿低温，限制因素是温度；中游地区低湿高温，水分条件是限制因素；下游地区低湿高温，在河岸带高盐碱，水分和土壤盐分是限制因素。

（3）由于生境条件不同，草地的生产力有极大的差异。而气温和降水存在着年际变化，草地的生产力也将表现出年际的变化。黑河流域植被对气候变化有着极大的依赖性和敏感性，在未来气候变化的背景下，草地的类型和生产能力也将相应发生变化，草地管理如何应对气候变化，将是管理者面临的极大挑战。

（4）流域草地潜在模型的建立，不仅实现了流域潜在草地分布可视化，而且获得草原潜在分布和实际分布之间的关系。潜在分布可作为背景，分析不同时期实际草地分布的变化。

（5）草地潜在分布的生物地理模型的变量与气候模型的输出结果一致，将该生物地理模型与气候模式耦合，可获得草原分布与自然驱动力的动力学关系，能够分析未来气候变化情景下草地的动态变化。

（6）潜在分布可作为草地管理的依据。利用潜在分布可以分析现实草地分布区不合理的扩张与缩减，如在上游牧民对林灌地的利用将草地扩展到林地或灌木林的分布区，在中下游通过利用地下和地表水将荒漠或戈壁开垦成草地，这将导致草地分布区的扩大，但是扩大的区域将是不稳的，一旦人文因素消失，草地将消失，原来的植被景观将重现。但是如果草地利用不合理，下游的草地将会荒漠化，上游的草地将随着水土的流失而减少，草地分布区将萎缩，草地的消失将不可逆转。

第5章 黑河流域近30年来草地变化分析

草地是黑河流域重要的生态系统，也是该流域发展畜牧业的物质基础，深受自然和人为活动的双重影响。自然条件的变化和人类活动的干扰，将使草地的类型及其覆盖度发生变化，分析草地的变化及变化的驱动力有利于草地的管理和维护，使其保持可持续发展。目前对于大范围草地变化的研究多采用遥感手段，利用多期美国陆地资源卫星LandsatTM影像数据，分析不同时期草地的空间分布和类型转换，是理解草地变化驱动力的首要条件，也为草地的潜在分布与现实分布的关系建立提供数据基础。

5.1 数据的获取与方法

下载1985年、2000年和2013年美国陆地资源卫星Landsat-5TM和Landsat-7ETM+影像数据，收集黑河流域90 m分辨率的DEM、黑河流域草场类型图、土地类型图、土壤图和土地资源评价图等图件。

应用遥感图像处理软件ERDAS IMAGINE，采用人机交互的判读分析方法，解译出黑河流域的草地类型图。根据草地类型图对解译的结果进行修正。应用GIS空间分析技术，统计草地与其他土地利用类型和草地类型间的转换矩阵，分析不同时期草地与其他类型的转换关系，阐明变化的驱动力。基于草地类型，查询相关研究成果，分析不同草地类型的生态承载力。

5.2 结果

5.2.1 黑河流域不同时期草地的分布格局

黑河流域上、中、下游隶属不同的行政区，以县域为单元，分析草地近 30 年的变化。利用遥感影像解译出的黑河流域草地分布图见图 5-1～图 5-3。从图中可以看出，黑河下游草地除河流两岸有中覆盖草地外，区域以低覆盖草地为主，低覆盖草地主要分布在东部的古日乃和西部的马鬃山；中游除山丹、民乐有高覆盖草地外，其他部分均为低覆盖草地，草地面积较小；上游高、中、低覆盖草地并存，草地面积较大。

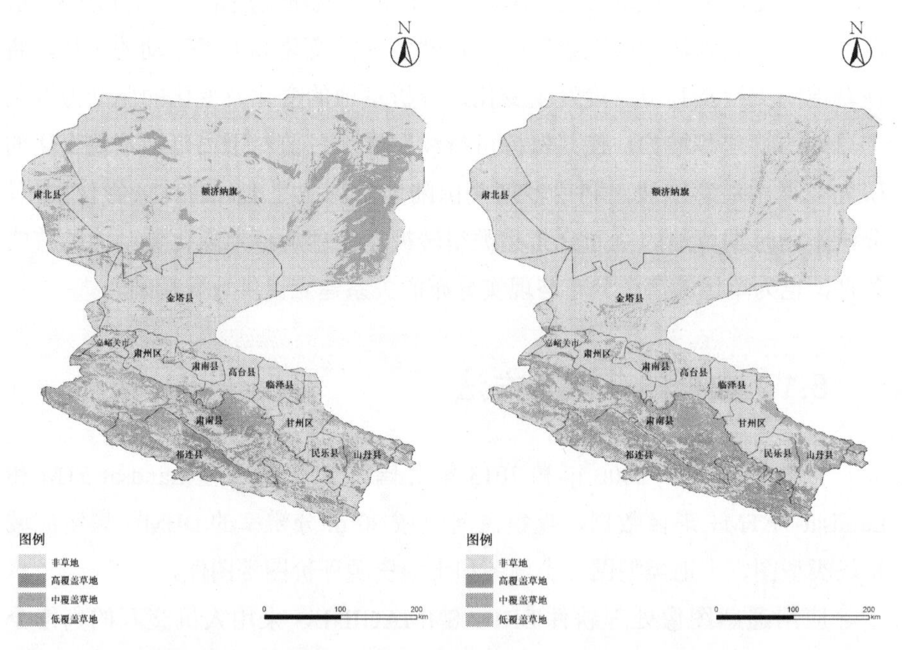

图 5-1　黑河流域草地空间分布（1985 年）（文后附彩图）

图 5-2　黑河流域草地空间分布（2000 年）（文后附彩图）

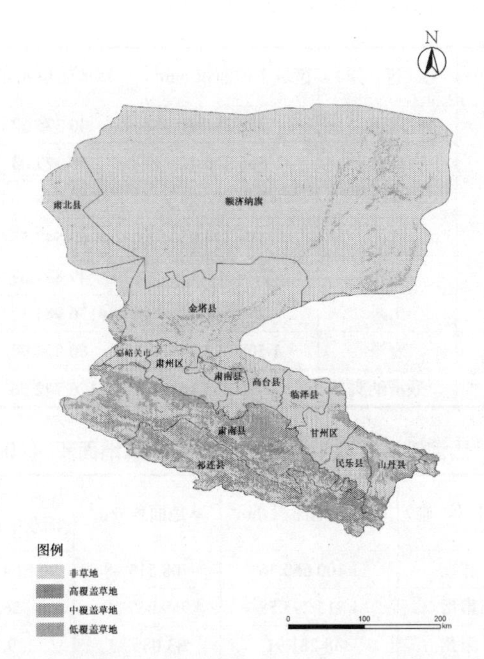

图 5-3　黑河流域草地空间分布（2013 年）（文后附彩图）

5.2.2　黑河流域草地近 30 年的空间变化分析

将 1985 ～ 2013 年划分为两个阶段，即 1985 ～ 2000 年和 2000 ～ 2013 年，不同时期草地占国土面积的比例见表 5-1～表 5-3。由表可知，在第一阶段，上游草地明显增加，增加幅度为 9% 左右；在中游肃州、高台、临泽和甘州明显减少，减少的幅度在 2% ～ 6.6%，其他区域略有增加；下游额济纳旗大幅度减少，减少幅度在 14.21%。第二阶段，上游的祁连县略有减少，肃南县增加了 1.67%；中游除甘州略有增加外，其余县均有不同程度的减少，减少最多的是肃州区，减少了 2.73%，其次是民乐，减少了 1.14%；下游略有增加。

表5-1　黑河流域行政划分及国土面积和草地面积(1985年)

黑河流域	省（自治区）	县（区、旗）	区域土地面积 /hm^2	草地面积 /hm^2	草地占区域土地面积的比例 /%
上游	青海	祁连	1 100 660.46	604 516.06	54.92
	甘肃	肃南	1 715 295.58	818 377.54	47.71

续表

黑河流域	省（自治区）	县（区、旗）	区域土地面积 /hm^2	草地面积 /hm^2	草地占区域土地面积的比例 /%
中游	甘肃	肃州	338 281.91	40 375.52	11.94
		高台	366 576.47	58 973.03	16.09
		临泽	246 016.29	32 085.80	13.04
		甘州	319 374.05	65 547.55	20.52
		民乐	224 505.13	49 837.67	22.20
		山丹	299 686.51	136 984.32	45.71
下游	甘肃	金塔	1 508 327.37	60 904.90	4.04
	内蒙古	额济纳旗	6 996 865.16	1 176 742.56	16.82

表5-2 黑河流域行政划分及国土面积和草地面积（2000年）

黑河流域	省（自治区）	县（区、旗）	区域土地面积 /hm^2	草地面积 /hm^2	草地占区域土地面积的比例 /%	1985 ~ 2000 年草地变化 /%
上游	青海	祁连	1 100 660.46	708 513.48	64.37	9.45
	甘肃	肃南	1 715 295.58	969 067.39	56.50	8.79
中游	甘肃	肃州	338 281.91	33 089.83	9.78	−2.16
		高台	366 576.47	37 025.96	10.10	−5.99
		临泽	246 016.29	15 849.39	6.44	−6.60
		甘州	319 374.05	46 164.23	14.45	−6.07
		民乐	224 505.13	56 777.35	25.29	3.09
		山丹	299 686.51	154 668.25	51.61	5.90
下游	甘肃	金塔	1 508 327.37	69 277.73	4.59	0.55
	内蒙古	额济纳旗	6 996 865.16	182 803.50	2.61	−14.21

表5-3 黑河流域行政划分及国土面积和草地面积（2013年）

黑河流域	省(自治区)	县（区、旗）	区域土地面积 /hm^2	草地面积 /hm^2	草地占区域土地面积的比例 / %	2000 ~ 2013 年草地变化 /%
上游	青海	祁连	1 100 660.46	702 835.11	63.86	−0.51
	甘肃	肃南	1 715 295.58	997 717.83	58.17	1.67
中游	甘肃	肃州	338 281.91	23 849.16	7.05	−2.73
		高台	366 576.47	33 225.42	9.06	−1.04
		临泽	246 016.29	13 583.41	5.52	−0.92
		甘州	319 374.05	46 564.02	14.58	0.13
		民乐	224 505.13	54 216.13	24.15	−1.14
		山丹	299 686.51	152 392.49	50.85	−0.76
下游	甘肃	金塔	1 508 327.37	75 125.50	4.98	0.39
	内蒙古	额济纳旗	6 996 865.16	209 089.37	2.99	0.38

5.2.3 各县草地近30年的空间变化分析

草地的变化主要发生在草地与其他土地利用类型或草地间的相互转化，以下分析各县区两个时间段草地与其他土地利用类型的转化状况。

1. 祁连县草地近30年的空间变化分析

祁连县位于黑河流域的上游，是黑河流域主要的草地分布区，草地分别占国土面积的54.92%（1985年）、64.37%（2000年）和63.86%（2013年），1985～2000年草地增加近10%，2000～2013年草地面积保持稳定。1985～2000年草地与其他类型的转换矩阵见表5-4。由表5-4可以看出：草地的增加主要由灌木林转换为草地，草地的减少主要是由草地转换为沼泽和高寒苔原，草地中不同的类型也发生了转换，尤其中覆盖度的草地向低覆盖草地的转换显著。2000～2013年草地与其他类型的转换矩阵见表5-5。由表5-5可以看出：高覆盖草地与沼泽转换显著，由沼泽转换为高覆盖草地大于高覆盖草地转换为沼泽草地，低覆盖草地转换成高寒苔原的面积也较大。此外，草地中不同的类型也发生了转换，高覆盖草地向中覆盖度的草地和中覆盖草地向低覆盖草地的转换地面积较大，表现出草地由向退化的方向发展。

表5-4 祁连县1985～2000年草地与其他土地利用类型的转换矩阵（单位：km^2）

1985年 \ 2000年	高覆盖草地	中覆盖草地	低覆盖草地	沼泽	高寒苔原
灌木林地	685.14	179.44	168.14		
高覆盖草地	1022.06	410.40	338.54	242.77	
中覆盖草地	537.82	487.16	548.74	186.81	151.02
低覆盖草地	309.15	336.07	723.91	109.08	267.59

注：只统计面积大于100 km^2的转换类型

表5-5 祁连县2000～2013年草地与其他土地利用类型的转换矩阵（单位：km^2）

2000年 \ 2013年	高覆盖草地	中覆盖草地	低覆盖草地	戈壁	沼泽	高寒苔原
高覆盖草地	2902.56	91.68	27.50		45.06	
中覆盖草地	33.89	1522.24	78.25			
低覆盖草地			2149.28	35.42		81.02
沼泽	89.01					

注：只统计面积大于30 km^2的转换类型

2. 肃南县草地近 30 年的空间变化分析

肃南县位于黑河流域的上游，从分布面积上看，草地是黑河流域的第一大草地分布区，分别占国土面积的 47.71%（1985 年）、56.50%（2000 年）和 58.17%（2013 年），1985 ～ 2000 年草地增加近 9%，2000 ～ 2013 年草地面积略有增加。1985 ～ 2000 年草地与其他类型的转换矩阵见表 5-6。由表 5-6 可以看出：草地的增加主要是由灌木林地、疏林地、盐碱地、裸岩和高寒苔原向各类草地转换的结果，草地的减少是由各类草地向河渠、河滩地、戈壁和高寒苔原进行转换的结果，草地中不同的类型也发生了转换，其中向低覆盖草地转化的面积较大。草地向退化的方向发展，尤其高、中覆盖度的草地向低覆盖草地转换的面积大于反方向转换的面积。2000 ～ 2013 年草地与其他类型的转换矩阵见表 5-7。由表 5-7 可以看出：草地面积略有增加，戈壁、沼泽、高寒苔原和平原旱地向低覆盖草地转换的面积较大，高海拔区域低覆盖草地有一定的面积转换成了高寒苔原，在低海拔区域，低覆盖草地也有一定的面积转换成了戈壁。草地中不同的类型也发生了转换，比较分析发现，草地向良好的方向发展。

表5-6　肃南县1985～2000年草地与其他土地利用类型的转换矩阵（单位：km^2）

1985 年 \ 2000 年	高覆盖草地	中覆盖草地	低覆盖草地	河渠	河滩地	戈壁	高寒苔原
有林地	356.14	234.32	339.17				
灌木林地	280.71	160.91	210.33				
疏林地			122.92				
高覆盖草地	434.18	331.49	686.41				
中覆盖草地	191.02	593.35	1228.44	100.64	65.23	442.61	105.14
低覆盖草地	213.64	607.30	1441.79	105.87	89.86	257.06	681.17
盐碱地			113.89				
裸岩	116.60	334.23	870.96				
高寒苔原	155.94		306.92				

注：只统计面积大于 50 km^2 的转换类型

表5-7　肃南县2000～2013年草地与其他土地利用类型的转换矩阵（单位：km^2）

2000 年 \ 2013 年	高覆盖草地	中覆盖草地	低覆盖草地	戈壁	高寒苔原
高覆盖草地	1606.32	30.75	132.47		
中覆盖草地	54.85	2311.45	62.02		

续表

2000年＼2013年	高覆盖草地	中覆盖草地	低覆盖草地	戈壁	高寒苔原
低覆盖草地	124.78	82.07	5094.26	31.30	66.27
戈壁			57.18		
沼泽			53.24		
高寒苔原		32.97	67.97		
平原旱地			52.60		

注：只统计面积大于 30 km^2 的转换类型

3. 肃州区草地近30年的空间变化分析

肃州区属于酒泉地区，位于黑河流域的中游的西部。降雨稀少，气候干旱，草地以低覆盖类型为主。草地分布面积在1985年、2000年和2013年分别为40 375.52 hm^2、33 089.83 hm^{2}和23 849.16 hm^2，分别占区域土地面积的11.94%、9.78%和7.05%。1985～2000年草地面积减少了18.04%，2000～2013年，草地面积减少了27.93%。1985～2000年草地与其他类型的转换矩阵见表5-8。由表5-8可以看出：低覆盖草地向戈壁、裸土和平原旱地转换，向戈壁转换的面积最大。2000～2013年草地与其他类型的转换矩阵见表5-9。由表5-9可以看出：低覆盖草地面主要向平原旱地转换。

表5-8　肃州区1985～2000年草地与其他土地利用类型的转换矩阵（单位：km^2）

1985年＼2000年	中覆盖草地	低覆盖草地	戈壁	裸土	平原旱地
低覆盖草地	22.60	89.61	147.64	27.35	75.89
戈壁		59.15			
盐碱		95.08			

注：只统计面积大于 20 km^2 的转换类型

表5-9　肃州区2000～2013年草地与其他土地利用类型的转换矩阵（单位：km^2）

2000年＼2013年	中覆盖草地	低覆盖草地	戈壁	平原旱地
中覆盖草地	27.20			
低覆盖草地		143.52	24.20	62.24
戈壁		31.40		

注：只统计面积大于 20 km^2 的转换类型

4. 高台县草地近30年的空间变化分析

高台县位于黑河流域的中游的西部，降雨稀少，气候干旱，草地以低覆盖类型为主。草地分布面积在1985年、2000年和2013年分别为58 973.03 hm^2，37 025.96 hm^2 和33 225.42 hm^2，分别占区域土地面积的16.09%、10.10%和9.06%。1985～2000年草地面积减少了37.22%，2000～2013年，草地面积减少了10.26%。1985～2000年草地与其他类型的转换矩阵见表5-10。由表5-10可以看出：低覆盖草地向河滩地、戈壁、沼泽和平原旱地转换，其中向戈壁转换的面积最大；沙地、戈壁和盐碱地向低覆盖草地转换，其中盐碱地向草地转换的面积最大。2000～2013年草地与其他类型的转换矩阵见表5-11。由表5-11可以看出：相比1985～2000年，2000～2013年草地的面积减少幅度降低，低覆盖草地面向沙地转换33.00 km^2，向盐碱地转换11.15 km^2，向平原旱地转换17.81 km^2。

表5-10 高台县1985～2000年草地与其他土地利用类型的转换矩阵（单位：km^2）

1985年 \ 2000年	中覆盖草地	低覆盖草地	河滩地	沙地	戈壁	沼泽	平原旱地
中覆盖草地		26.06					
低覆盖草地	26.51	114.00	28.57		243.56	41.92	18.73
沙地		37.70					
戈壁		26.68					
盐碱地		71.31					

注：只统计面积大于10 km^2 的转换类型

表5-11 高台县2000～2013年草地与其他土地利用类型的转换矩阵（单位：km^2）

2000年 \ 2013年	中覆盖草地	低覆盖草地	沙地	盐碱地	平原旱地
中覆盖草地	44.28				
低覆盖草地		216.83	33.00	11.15	17.81

注：只统计面积大于10 km^2 的转换类型

5. 临泽县草地近30年的空间变化分析

临泽县位于黑河流域的中游的西部，是流域国土面积较小的一个县，为农业区。降雨稀少，气候干旱，草地以低覆盖类型为主。草地分布面积在1985年、2000年和2013年分别为32 085.80 hm^2、15 849.39 hm^2 和13 583.41 hm^2，分别占区域土地面积的13.04%、6.44%和5.52%。1985～2000年草地面积减少了一半（50.60%），2000～2013年草地面积

减少了14.30%。1985～2000年草地与其他类型的转换矩阵见表5-12。由表5-12可以看出：低覆盖草地向沙地、戈壁、裸岩和平原旱地转换，其中向平原旱地转换的面积最大，其次为戈壁；盐碱地向低覆盖草地转换了15.55 km^2。2000～2013年草地与其他类型的转换矩阵见表5-13。由表5-13可以看出：相比1985～2000年，2000～2013年草地的面积减少幅度降低，低覆盖草地主要向平原旱地转换。

表5-12 临泽县1985～2000年草地与其他土地利用类型的转换矩阵（单位：km^2）

1985年＼2000年	中覆盖草地	低覆盖草地	沙地	戈壁	裸岩	平原旱地
中覆盖草地		7.23				11.67
低覆盖草地	6.27	49.95	19.78	55.78	51.23	90.72
沙地		23.17				
戈壁		21.58				
盐碱地	7.34	15.55				

注：只统计面积大于7 km^{2}的转换类型

表5-13 临泽县2000～2013年草地与其他土地利用类型的转换矩阵（单位：km^2）

2000年＼2013年	疏林地	高覆盖草地	中覆盖草地	低覆盖草地	平原旱地
中覆盖草地			10.60		
低覆盖草地	15.68			85.22	20.76
沙地				16.82	
戈壁				13.12	

注：只统计面积大于10 km^{2}的转换类型

6. 甘州区草地近30年的空间变化分析

甘州区位于黑河流域的中游的中部，属于张掖市的一个区，由于拥有灌溉的优势，为典型的农业区。降雨稀少，气候干旱，草地以低覆盖类型为主。草地分布面积在1985年、2000年和2013年分别为65 547.55 hm^2、46 164.23 hm^{2}和46 564.02 hm^2，分别占区域土地面积的20.52%、14.45%和14.58%。1985～2000年草地面积减少了29.57%，2000～2013年草地基本上保持不变。1985～2000年草地与其他类型的转换矩阵见表5-14。由表5-14可以看出：低覆盖草地主要向戈壁和平原旱地转换，向戈壁转换的面积最大，其次为平原旱地；裸岩、戈壁和河滩地不同程度地向低覆盖草地转换。2000～2013年草地与其他类型的转换矩阵见表5-15。由表5-15可以看出：相比1985～2000年，2000～2013年草地的面积保持稳定，低覆盖草地主要向平原旱地转换，同时沙地和戈壁向低覆盖草地转换。

表5-14 甘州区1985～2000年草地与其他土地利用类型的转换矩阵（单位：km^2）

1985年＼2000年	中覆盖草地	低覆盖草地	河渠	河滩地	沙地	戈壁	裸岩	平原旱地
高覆盖草地		10.99						
中覆盖草地	8.46	32.70						14.58
低覆盖草地	35.53	108.55	11.28	9.30	14.75	190.20	27.17	122.89
河滩地	6.78	15.14						
沙地		9.02						
戈壁	14.78	46.63						
裸岩	47.83	94.23						

注：只统计面积大于6 km^2的转换类型

表5-15 甘州区2000～2013年草地与其他土地利用类型的转换矩阵（单位：km^2）

2000年＼2013年	中覆盖草地	低覆盖草地	平原旱地
中覆盖草地	117.49		
低覆盖草地		305.09	18.58
河渠			
沙地		11.08	
戈壁	9.00	13.42	

注：只统计面积大于9 km^2的转换类型

7. 民乐县草地近30年的空间变化分析

民乐县位于黑河流域的中游的东部，属于平原向山区过度的一个县。由于海拔的增加，降雨相对增加，高、中和低覆盖草地类型俱全，是黑河流域农牧交错的区域。草地分布面积在1985年、2000年和2013年分别为49 837.67 hm^2、56 777.35 hm^2和54 216.13 hm^2，分别占区域土地面积的22.20%、25.29%和24.15%。1985～2000年草地面积增加了13.92%。2000～2013年草地略有减少，减少了4.51%。1985～2000年草地与其他类型的转换矩阵见表5-16。由表5-16可以看出：草地的增加主要由林地转化而来；同时草地也向河渠、沙地、戈壁、裸土、丘陵旱地和平原旱地不同程度地转换。2000～2013年草地与其他类型的转换矩阵见表5-17。由

表5-17可以看出：相比1985～2000年，2000～2013年草地的面积保持稳定，草地内部不同类型在相互转换。

表5-16　民乐县1985～2000年草地与其他土地利用类型的转换矩阵（单位：km²）

1985年 \ 2000年	高覆盖草地	中覆盖草地	低覆盖草地	河渠	沙地	戈壁	裸土	丘陵旱地	平原旱地
有林地	29.82								
灌木林地	95.52	32.46							
疏林地	10.24								
高覆盖草地			14.91					10.19	
中覆盖草地	13.33	32.87	41.36					44.64	11.42
低覆盖草地		19.54	107.16	9.02	11.36	40.38	19.15	30.33	55.52
河滩地			11.77						
戈壁			23.36						
裸土			24.70						
裸岩	10.89								
平原旱地		8.38	29.08						

注：只统计面积大于10 km²的转换类型

表5-17　民乐县2000～2013年草地与其他土地利用类型的转换矩阵（单位：km²）

2000年 \ 2013年	高覆盖草地	中覆盖草地	低覆盖草地	平原旱地
高覆盖草地	144.95		15.46	
中覆盖草地		116.19		
低覆盖草地			226.67	40.21
丘陵旱地			8.98	

注：只统计面积大于9 km²的转换类型

8. 山丹县草地近30年的空间变化分析

山丹县位于黑河流域的中游的东部，土地面积较大，草地占区域土地面积的近一半。高、中和低覆盖草地类型俱全，是黑河流域农牧交错的区域。草地分布面积在1985年、2000年和2013年分别为136 984.32 hm²、154 668.25 hm²和152 392.49 hm²，分别占区域土地面积的45.71%、51.61%和50.85%。1985～2000年草地面积增加了12.91%。2000～2013年草地略有减少，减少了1.47%。1985～2000年草地与其他类型的转换矩阵见表5-18。由表5-18可以看出：草地的增加主要由林地转换，河滩地、戈壁、裸土、丘陵旱地和平原旱地也转换为不同的草地类型；同时草地也向戈壁、裸土、裸岩、丘陵旱地和平原旱地不同程度地转换。2000 ～ 2013 年草地与其他类型的转换矩阵见表 5-19。由表 5-19 可以看出：草地内部不同类型在相互转换，草地面积保持稳定。

表5-18　山丹县1985～2000年草地与其他土地利用类型的转换矩阵（单位：km^2）

1985 年 \ 2000 年	高覆盖草地	中覆盖草地	低覆盖草地	河滩地	沙地	戈壁	裸土	裸岩	平原旱地
有林地	53.22	22.43	32.12						
疏林地	15.60	9.68	15.62						
高覆盖草地		36.03	101.78			18.45	13.38		
中覆盖草地		128.15	280.96			33.45	9.19	9.33	20.01
低覆盖草地		82.89	383.75	15.15	11.90	62.98	68.15		32.17
河滩地		14.66	27.00						
戈壁		35.09	160.06						
裸土			11.84						
丘陵旱地			30.40						
平原旱地			32.72						

注：只统计面积大于 9 km^2 的转换类型

表5-19　山丹县2000～2013年草地与其他土地利用类型的转换矩阵（单位：km^2）

2000 年 \ 2013 年	高覆盖草地	中覆盖草地	低覆盖草地	丘陵旱地	平原旱地
高覆盖草地	84.48				
中覆盖草地		341.75	7.07		
低覆盖草地			1053.34	7.01	22.37
平原旱地			11.13		

注：只统计面积大于 7 km^2 的转换类型

9. 金塔县草地近 30 年的空间变化分析

金塔县位于黑河流域下游的南部，草地占区域土地面积面积不足 5%，降水量少，气候非常干旱，以低覆盖草地类型为主。草地分布面积在 1985 年、2000 年和 2013 年分别为 60 904.90 hm^2，69 277.73 hm^2 和 75 125.50 hm^2，分别占区域土地面积的 4.04%、4.59% 和 4.98%。1985 ～ 2000 年草地面积增加了 13.75%，2000 ～ 2013 年草地面积增加了 8.44%。1985 ～ 2000 年草地与其他类型的转换矩阵见表 5-20。由表 5-20 可以看出：草地的增加主要由疏林地和灌木林地转换为低覆盖草地，河滩地、沙地、戈壁、盐碱地、裸土和平原旱地也转换为低覆盖草地；同时低覆盖草地也向沙地、戈壁、

裸岩和平原旱地不同程度地转换。2000～2013年草地与其他类型的转换矩阵见表5-21。由表5-21可以看出：草地的增加主要由沙地转换而成，同时，低覆盖草地也有一定的面积转换成了平原旱地和沙地。

表5-20　金塔县1985～2000年草地与其他土地利用类型的转换矩阵（单位：km^2）

1985年 \ 2000年	高覆盖草地	中覆盖草地	低覆盖草地	沙地	戈壁	裸岩	平原旱地
灌木林地			22.52				
疏林地			10.21				
低覆盖草地			175.52	154.71	98.35	80.27	27.82
河滩地			13.29				
沙地			135.80				
戈壁			86.98				
盐碱地			158.09				
裸土			16.52				
平原旱地			23.10				

注：只统计面积大于10 km^2的转换类型

表5-21　金塔县2000～2013年草地与其他土地利用类型的转换矩阵（单位：km^2）

2000年 \ 2013年	中覆盖草地	低覆盖草地	河渠	沙地	戈壁	盐碱	沼泽	裸土	平原旱地
疏林地		8.62							
中覆盖草地	14.34								
低覆盖草地	19.18	374.27	6.36	64.02	12.20	10.03	47.66	8.32	104.42
河滩地		8.07							
沙地		230.97							
戈壁		27.92							
盐碱地		6.90							
平原旱地		9.36							

注：只统计面积大于6 km^2的转换类型

10. 额济纳旗草地近30年的空间变化分析

额济纳旗位于黑河流域下游，区域土地面积是流域中最大的，该区域降水量稀少，气候极端干旱，低覆盖草地类型占主要优势。草地分布面积在1985年、2000年和2013年分别为1 176 742.56 hm^2、182 803.50 hm^2和209 089.37 hm^2，分别占区域土地面积的16.82%、2.61%和2.99%。

1985～2000年草地急剧减少，几乎减少了84.47%。2000～2013年，草地增加了14.38%。1985～2000年草地与其他类型的转换矩阵见表5-22。由表5-22可以看出：大面积低覆盖草地转换为沙地、戈壁、盐碱地和裸岩，河滩地、沙地、戈壁、盐碱地、裸土和平原旱地也转换为低覆盖草地；同时低覆盖草地也向沙地、戈壁、裸岩和平原旱地不同程度地转换。2000～2013年草地与其他类型的转换矩阵见表5-23。由表5-23可以看出：草地的增加主要由沙地转换而成。

表5-22　额济纳旗1985～2000年草地与其他土地利用类型的转换矩阵（单位：km^2）

1985年 \ 2000年	有林地	疏林地	高覆盖草地	中覆盖草地	低覆盖草地	河滩地	沙地	戈壁	盐碱地	裸岩
灌木林地			25.85	102.86	134.54					
中覆盖草地		25.27		75.12	97.63		74.98	47.74	16.09	
低覆盖草地	23.06	47.11		169.42	609.21	171.62	3822.39	5037.42	561.89	811.08
沙地				18.97	100.86					
戈壁				12.07	159.57					
盐碱地				37.17	168.19					

注：只统计面积大于20 km^2的转换类型

表5-23　额济纳旗2000～2013年草地与其他土地利用类型的转换矩阵（单位：km^2）

2000年 \ 2013年	疏林地	高覆盖草地	中覆盖草地	低覆盖草地	沙地	戈壁	盐碱地	平原旱地
高覆盖草地		36.32						
中覆盖草地	10.36		354.74	27.56	13.46			10.68
低覆盖草地	21.55		37.27	1033.18	105.11	50.49	28.73	18.67
沙地		32.58	97.91	247.60				
戈壁			14.58	63.46				
盐碱地			9.19	50.19				

注：只统计面积大于10 km^2的转换类型

5.2.4　草地承载力分析

草地载畜量被定义为特定时期内在草地资源不受破坏的条件下，一定面积的草地能够承载的家畜数量，可分为产量载畜量和营养载畜量。考虑产量载畜量，大多数研究以50%的牧草可利用率计算载畜量，扣除了避免土壤侵蚀和遭遇干旱时保证植被群落弹性和再生需要的牧草生物量，以及满足昆虫、土壤无脊椎动物及野生动物需要的牧草生物量。

草地承载力的计算如下：

$$C_i = \frac{F_i(1-E_i)(1-\mathrm{DW})(1-\mathrm{DS})}{G} \tag{5-1}$$

$$S = \sum_{i=1}^{n} C_i A_i \tag{5-2}$$

式中，C_i 为草地类型 i 的承载力（单位面积载畜量）；F_i 为草地类型 i 的单位产草量；E_i 为草地类型 i 的牧草生态生物量分配；DW 和 DS 分别为在不同的等级距离水源和所处坡度的载畜量减少比例；G 为羊的食草量；S 为载畜量；A_i 为草地类型 i 的面积；n 为草地类型的种类数。

黑河流域气候差异较大，草地类型复杂，牧草生物量差别很大。根据植被盖度，可将黑河流域草地分为高覆盖、中覆盖、低覆盖草地。结合黑河流域具体情况，分别将黑河流域高覆盖、中覆盖、低覆盖草地满足昆虫、土壤无脊椎动物及野生动物生态需要的牧草生物量占总牧草量的比例定为15%、25% 和 35%。DW 和 DS 根据表 5-24 进行调整。黑河流域山区系个体较大的改良细毛羊和藏系绵羊，日食量按 5 kg 计算；平原荒漠区蒙古羊个体较小，一般按 3 kg 计算；半农半牧和农区介于其间，一般按 4 kg 计算。根据野外调查，分布在不同区域的草地类型相应的产草量见表 5-25。

表5-24　黑河流域载畜量与坡度和距离水源的调整系统

距离水源远近 /km	载畜量减少比例 /%	坡度 /%	载畜量减少比例 /%
0 ～ 1.6	0	0 ～ 5	0
1.6 ～ 3.2	50	5 ～ 15	30
>3.2	100	15 ～ 25	50
		25 ～ 35	70
		>35	100

表5-25　黑河流域天然草场分类、分布及产草量

草地类型	分布区域	覆盖度 /%	产草量 / (kg/hm²)
低湿沼泽草场	中下游河滩地	高覆盖度（>75%）	6000 ～ 7500
高寒沼泽草场	上游河源和高山	高覆盖度（>75%）	375 ～ 450
低湿草甸草场	中下游河湖滩地	中高覆盖度（40% ～ 80%）	6000 ～ 5500
平原荒漠草场	中下游荒漠区	低覆盖度（<10%）	＜450
平原半荒漠草场	山麓和山前平原区	中低覆盖度（5% ～ 30%）	750 ～ 1200

续表

草地类型	分布区域	覆盖度 /%	产草量 /（kg/hm^2）
荒漠河岸林灌草丛草场	下游河岸疏林和灌草丛	高覆盖度（＞50%）	750 ～ 5700
山地荒漠草场	中下游低山丘陵区	低覆盖度（5% ～ 20%）	150 ～ 1050
山地半荒漠草场	上游低山和高山谷地	中覆盖度（30%）	750 ～ 1200
山地草原草场	海拔在 2100 ～ 3800 m	中高覆盖度（30% ～ 70%）	975 ～ 1425
山地草甸草场	海拔在 2800 ～ 3800 m	高覆盖度（＞75%）	1950 ～ 2250
山麓灌丛草甸草场	海拔在 3000 ～ 3700 m	高覆盖度（＞50%）	1350 ～ 2250

根据式（5-1）和式（5-2），利用前人研究成果、野外调查数据和草地的分布面积，对黑河流域承载力进行计算，得出黑河流域不同行政管理区域的草地承载量的空间分布（图 5-4～图 5-6）。

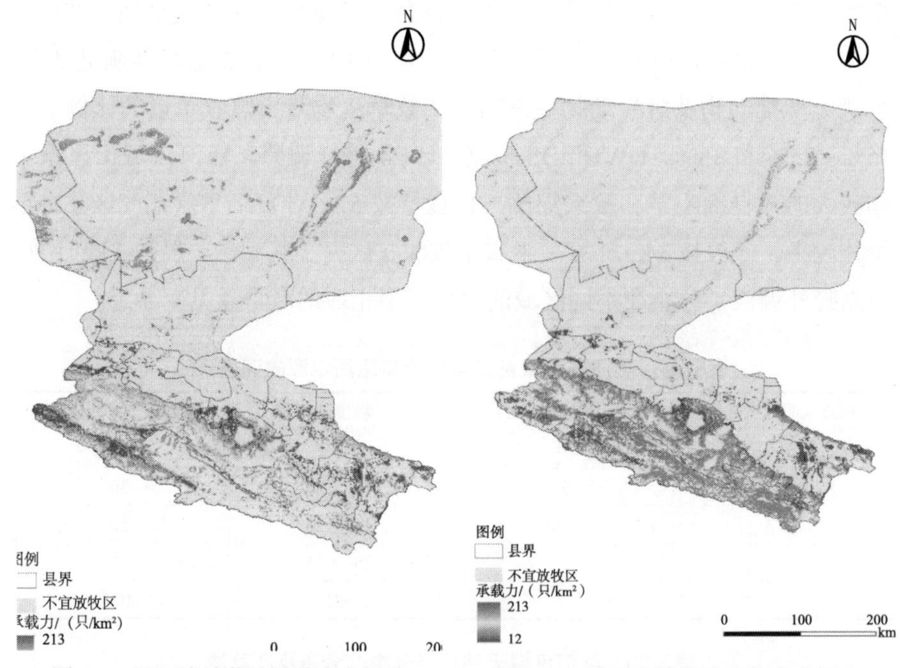

图 5-4　黑河流域草地承载力空间分布（1985 年）（文后附彩图）

图 5-5　黑河流域草地承载力空间分布（2000 年）（文后附彩图）

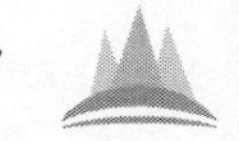

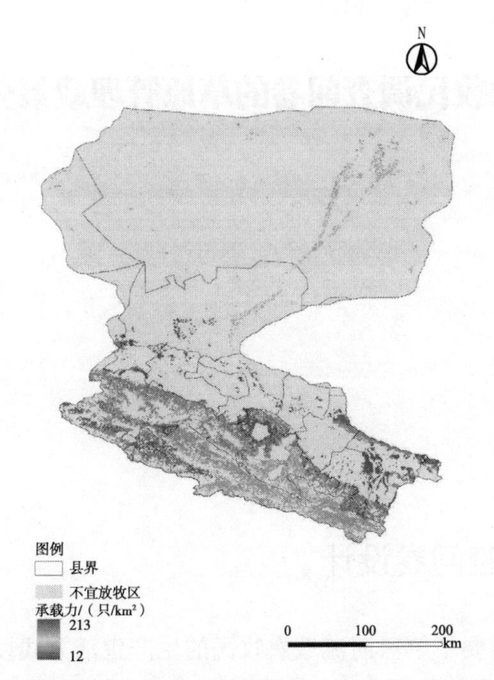

图 5-6　黑河流域草地承载力空间分布（2013 年）（文后附彩图）

5.3　结论

（1）利用遥感数据对不同时期黑河流域草地分类，分类结果显示，草地类型在近 30 年间有显著的变化。第一阶段（1985 ～ 2000 年），上游草地明显增加，中下游显著减少。第二阶段（2000 ～ 2013 年），整个流域草地变化不大。

（2）从草地与其他土地利用类型的转换矩阵看，草地的变化的驱动力以人为因素为主，政策的调整致使草地与其他土地利用类型相互转换显著。

（3）不同的自然条件组合，草地的生产力差异较大，草地的承载力不同。在草地管理中，应该严格按照草畜平衡的原则进行畜牧业的发展，过度放牧将会导致草地退化、草地资源不可持续利用。

（4）基于遥感数据解译的黑河流域草地分布格局，为潜在草地模拟和潜在植被与现实草地之间的关系分析奠定了基础。

第6章
基于农牧民调查问卷的草原管理政策分析

6.1 调查问卷设计

鉴于研究需要了解黑河流域农牧民的生产生活方式以及草原管理相关政策实施状况，有必要进行社会调查。本研究以黑河流域基层生产生活的农牧民为调查对象，以“农牧民关于黑河流域草原管理政策评价”为调查主题，设计“中国黑河流域草原管理政策研究——农牧民生产生活方式及政策调查”（附录1农牧民调查问卷），通过了解黑河流域草原管理最基层、最相关生产生活参与者的基本状况，获取真实信息，为本项研究及相关政策分析设计提供重要基础。

6.1.1 问卷内容

调查问卷以获得黑河流域相关基层农牧民生产生活方式及政策实施状况为目的，通过研究组成员讨论设计、相关领域专家的建议讨论分析，获得“中国黑河流域草原管理政策研究——农牧民生产生活方式及政策调查”问卷。调查问卷包括被调查人基本信息、被调查人的经济状况、黑河流域草原现状、黑河流域草原管理政策四个部分，每部分又设计多个题目进行黑河流域农牧民生产生活方式及草原管理政策相关内容的具体调查（具体问卷见附录1）。

第一部分为被调查人的基本信息，反映民族、文化等方面的差异，这

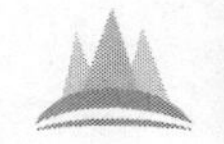

些差异会反映到被调查人对草原管理政策的认知感受方面。因此，在调查过程当中，问卷应该尽量能够按照民族比例发放，选择年龄较长、长期从事草原农牧业基础生产活动的受访者，同时保证性别、文化等的差异性与区分度，以保证调查结果的真实性、有效性与说服力。第二部分反映被调查人的经济状况，农牧民的基本经济状况是影响农牧民对草原政策态度的关键因素，通过对被调查者家庭的基本收入来源、收入数目、收入变化及支出状况等的统计，为后续农牧民的草原政策执行情况、执行效果及执行效果的原因分析提供分析基础。第三部分为黑河流域草原现状调查，这一部分反映草原政策执行以来草原生态环境变化的自然原因。第四部分为草原管理政策执行情况调查，包括黑河流域草原管理政策的执行情况、执行效果及产生这一效果的原因，并分析当前草原管理政策的问题所在。对黑河流域草原管理政策的执行情况、执行效果、效果原因及现存问题的讨论，可为未来草原管理政策的完善提供必要的现实支撑与数据支持（表 6-1）。

表6-1　农牧民调查问卷主要内容

调查方面	调查内容
被调查人基本信息	基本信息：民族、性别、年龄、文化程度
被调查人经济状况	家庭状况：家庭人口、劳动人口、家庭耐用品 家庭收入：畜牧业状况及收入、种植业状况及收入、主要收入来源、年均纯收入、收入变化 家庭支出：牧业支出、种植业支出、生活消费支出
黑河流域草原现状	气温变化、降水变化、草地生态变化原因
黑河流域草原管理政策	草原生态保护政策评价 草原生态保护补偿奖励情况 草原保护政策或措施的整体治理效果 草原破坏的处罚力度 违反草原管理政策的主要原因 草原保护政策或措施对收入的影响 草原保护政策或措施对生态环境的影响 农牧民提供收入的可行性措施 政府未来在草原管理方面的工作重点

6.1.2　调查方案

为保证调查结果的真实性、有效性，本项目采用实地调查方式。由来自中国科学院科技政策与管理科学研究所、兰州大学的本课题组相关研究人员通过深入黑河流域的相关地区，与从事农牧业基层生产的农牧民进行面对面的交流询问，获取与研究相关的一手数据资料，保证了调查结果的有效性与说服力。

研究调查覆盖黑河流域的全部县（区、旗），覆盖区域广泛，调查数据真实可靠。问卷一共发放1500份，回收问卷1150份，有效问卷1017份，其中甘肃省754份、内蒙古自治区148份、青海省115份，问卷有效率为88.4%。从具体的黑河流域各县（区、旗）分布来说，内蒙古额济纳旗148份，青海省祁连县115份，甘肃省高台县112份、金塔县149份、临泽县141份、民乐县109份、山丹县87份、肃南县138份、甘州区18份。除张掖市辖区甘州区以外的所有黑河流域县区问卷数均处于87～148份，占问卷总数的8%以上。调查问卷地区分布合理、均匀，涉及黑河流域所有县（区、旗）的基层农牧业生产劳动者，能够真实反映黑河流域草原管理的现实状况（表6-2）。

表6-2 农牧民调查问卷县（区、旗）分布

省份	地区	县（区、旗）	有效问卷数目
青海	海北藏族自治州	祁连县	115
甘肃	张掖市	肃南县	138
		山丹县	87
		民乐县	109
		甘州区	18
		临泽县	141
		高台县	112
	酒泉市	金塔县	149
内蒙古	阿拉善盟	额济纳旗	148

6.2 调查问卷结果

6.2.1 调查对象的基本特征

问卷调查的对象是黑河流域县（区、旗）的农牧民。从民族特征来看，以汉族为主，藏族、蒙古族、裕固族也占有一定的比例。从性别来看，男女比例大致为3∶1，男性居多。从年龄来看，被调查人年龄集中分布于21～60岁，其中31～50岁的青壮年劳动力占比60%以上。从文化程度来看，以小学到高中（中专）的受教育程度为主。综合被调查对象的民族、

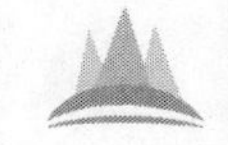

性别比例、年龄构成、文化程度水平，调查对象特征分布合理均匀，问卷能够真实反映研究需要探索的问题。

6.2.2 被调查人家庭经济状况

1. 家庭基本状况

家庭基本状况包括家庭劳动人口、家庭耐用品数量，家庭耐用品分为交通工具耐用品、生产工具耐用品及生活耐用品，其中交通工具耐用品包括电动车、摩托车、汽车，生产工具耐用品包括三轮车、四轮车、拖拉机、水泵，生活耐用品包括冰箱、电视机、洗衣机。家庭人口方面，通过图 6-1 数据可得，受访农牧民家庭平均人口接近 4 人，劳动力人口接近 2.5 人，其中从事农业生产 2 人，从事牧业生产 1.8 人，外出打工 1 人。总体来说，农牧民家庭劳动人口中，农业生产人口大于牧业生产人口，并且平均每个家庭就有 1 人外出打工。家庭拥有耐用品方面（图 6-2），通过对数据基本统计可得，交通工具耐用品、生产工具耐用品家庭拥有比例占比不高，生活耐用品家庭拥有比例占比很高，家庭拥有比例为“生活耐用品＞交通工具耐用品＞生产工具耐用品”。这说明现在当地农牧民家庭生活条件明显改善，生活水平显著提高，但是生产条件、交通通信条件依然相对落后，生产水平还有待提高。

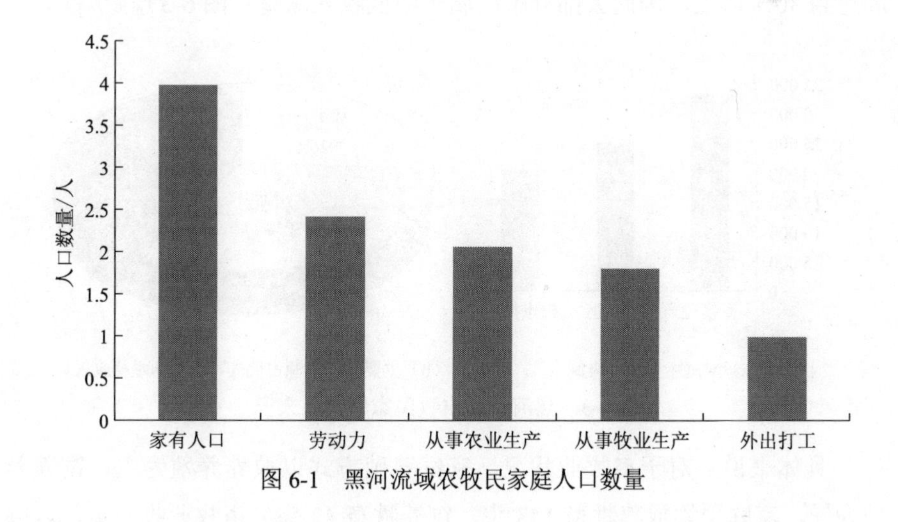

图 6-1　黑河流域农牧民家庭人口数量

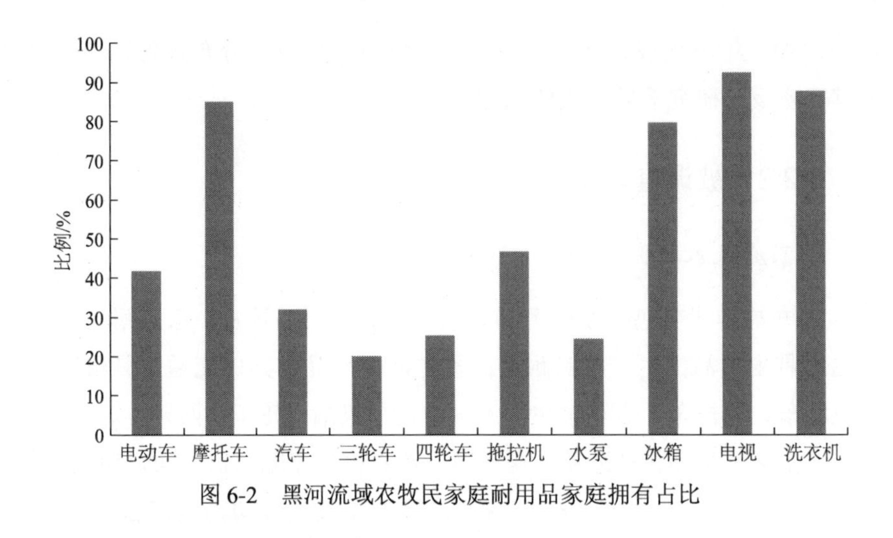

图 6-2　黑河流域农牧民家庭耐用品家庭拥有占比

2. 家庭收入状况

问卷结果显示，黑河流域农牧民家庭收入主要包括三部分：畜牧业收入、种植业收入及外出打工收入，其中家庭平均畜牧业收入为 33 310 元、种植业收入为 24 968 元、外出打工收入为 20 686 元。因此家庭“畜牧业收入＞种植业收入＞外出打工收入”，畜牧业收入成为家庭最主要的收入来源，打工收入逐渐成为家庭重要的经济来源渠道。对于家庭年均纯收入，3/4 的家庭年均纯收入介于 1 万～ 7 万元，其中 3 万～ 5 万元的家庭占比最高达到 30% 以上，因此大部分家庭属于中低收入家庭（图 6-3）。

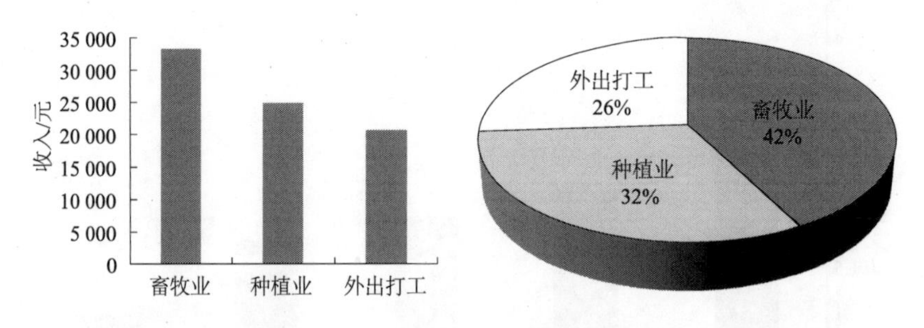

（a）家庭主要收入来源　　（b）主要收入来源占比（不包括非主要收入）

图 6-3　黑河流域农牧民家庭收入来源

具体来说：对于畜牧业状况，家庭养殖方式以放养养殖为主、饲养养殖为辅，家庭平均放养牲畜 143 头、饲养牲畜 42 头；畜牧业收入中，以肉

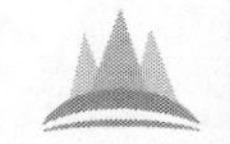

产品收入为主，家庭平均达到19 926元，毛、乳、皮收入为辅。对于种植业，家庭平均拥有耕地22.65亩，均亩产达到1400斤，大致收入25 000元。对于外出打工方面，平均每个家庭有一人外出打工，打工时间平均为150天，打工收入家均20 686元。三者构成黑河流域农牧民家庭收入的主要来源。

3. 家庭支出状况

家庭支出主要包括三个部分，牧业生产支出、种植业生产支出、家庭生活消费支出。综合计算得出，家庭总支出平均大致为95 759元，其中牧业平均支出为48 122元、种植业平均支出为22 431元、生活消费平均支出为25 206元，家庭牧业支出＞家庭生活消费支出＞家庭种植业支出。畜牧业支出中，购买幼畜、草料等基本的成本支出构成畜牧业支出的主要部分，分别达到16 708元、9025元，其他支出相对较少；农业支出中，雇工支出、化肥支出最高，分别达到5653元、4287元。

总的来说，家庭支出中畜牧业支出最高，占据家庭支出的一半左右，说明畜牧业生产成本较高，特别是购买幼畜支出，但畜牧业劳动力的投入较低。种植业成本投入较低，但劳动力的投入却更高。家庭生活消费支出总额较高，高于家庭种植业支出，说明当地农牧民家庭消费能力提高，生活水平改善；但食品支出占比较高也说明家庭恩格尔系数较高，农牧民生活水平改善有限；家庭文化教育支出较高，说明农牧民家庭普遍重视子女教育问题，也说明我国教育收费水平较高。

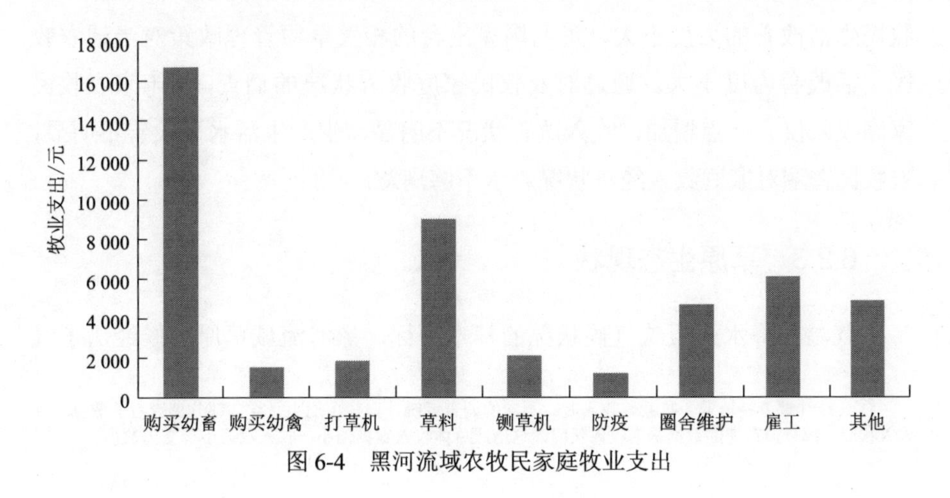

图6-4　黑河流域农牧民家庭牧业支出

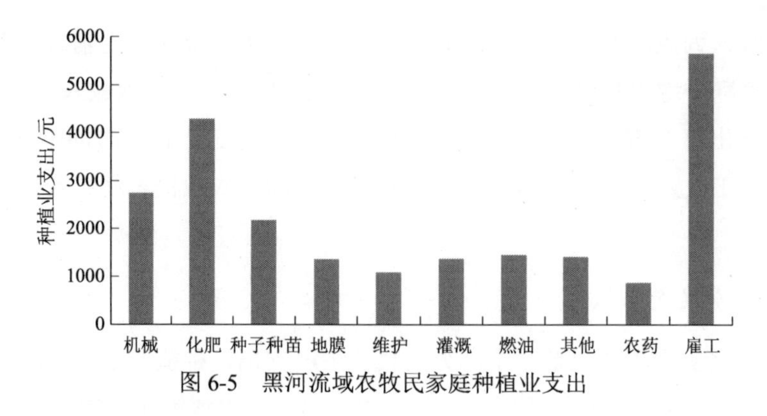

图 6-5　黑河流域农牧民家庭种植业支出

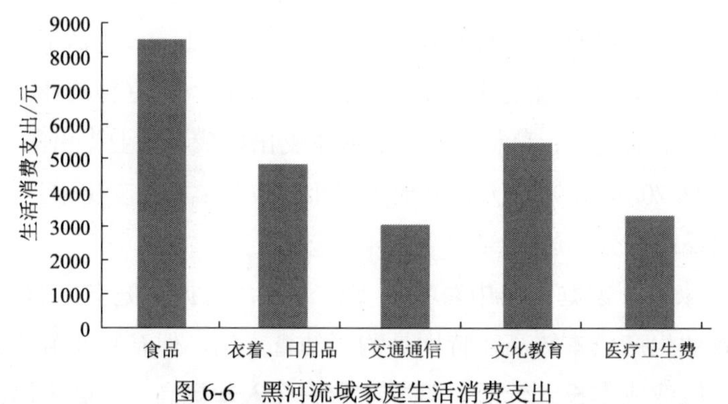

图 6-6　黑河流域家庭生活消费支出

对家庭收入、支出状况进行综合分析，家庭平均总收入大致为 78 964 元（没有考虑非主要收入），家庭平均总支出大致为 95 761 元，家庭平均纯收入处于 1 万～7 万元[①]。综合来说，黑河流域农牧民家庭支出数额较高，家庭收入仅能满足正常的家庭生产生活支出，很难有收支盈余，近年来农牧民生活改善的力度不大，说明国家出台的相关草原管理政策对当地农牧民生活改善力度不大。通过对农牧民家庭收入状况的调查，5 年来农牧民家庭收入仅有一点增加，收入改善状况不明显，生产生活水平改善不明显，农牧民普遍对家庭收入经济状况表示不够满意。

6.2.3　草原生态现状

气温与降水是反映气候状况的基本指标，黑河流域草原气候经历了气

① 统计调查种植业、牧业等收入时，不同的调查者统计口径可能不一样，部分调查者的数据是纯收入状况，部分调查者的数据是毛收入状况，这造成总体收入数据偏小，收入数据小于支出数据。

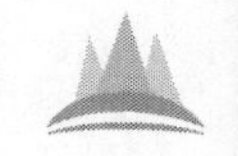

温升高、降水减少的变化过程。同时气温、降水是影响黑河流域草原环境的两个主要自然因素。通过对黑河流域气温、降水状况的统计，获得农牧民关于大致10年来流域自然环境因素变化的感知。关于气温状况，大致有60%的农牧民认为气温有一点升高；对于降水状况，大致35%的农牧民认为降水有一点减少。根据高宇（2013）的相关研究，近50年以来黑河流域的确存在气温显著升高、降水轻微减少的气候变化趋势，这验证了本研究的调查结果（表6-3）。

表6-3 黑河流域草原生态环境变化的自然原因

原因	选项	频数	相对频率/%
气温变化	有很大升高	140	13.8
	有一点升高	603	59.3
	基本不变	122	12.0
	有一点降低	84	8.3
	有很大降低	13	1.3
降水变化	有很大增加	18	1.8
	有一点增加	236	23.2
	基本没变	194	19.1
	有一点减少	353	34.7
	有很大减少	157	15.4

综合来说，气温升高、降水减少是影响黑河流域草原生态环境变化的主要自然因素，其他影响草原生态环境变化的原因可能包括水资源、土地、旱灾、暴雨、冻灾、人口等因素。其中，水资源减少、土地减少、旱灾增加、暴雨减少、冻灾增加、人口增加，也是可能影响黑河流域草地生态环境变化的自然或人文因素。

6.2.4 草原管理政策评价

1. 草原管理政策执行情况

草原管理政策包括退牧还草政策，退耕还林政策，禁牧、休牧、轮牧政策，草畜平衡政策，生态补偿政策等五大类；而生态补偿政策包括禁牧补助、退牧补助、退耕补助、草畜平衡奖励、人工草地补助、畜牧良种补贴、牧草良种补贴、生产资料综合补贴。综合来说，各个政策均得到较好的贯彻执行，特别是关于草原生态保护补助及奖励的政策得到很好的贯彻执行。黑河流域农牧民家庭平均每年获得政府补贴 14 259 元，并且基本能够按时、足额发放，说明政府在贯彻落实草原管理政策的同时重视农牧民生态补贴与补偿，奖励补偿政策贯彻落实比较完善。

2. 草原管理政策执行效果

对黑河流域草原管理政策执行效果进行调查分析，37.5% 的受访农牧民认为草原管理政策的整体效果比较好，因此总体来说政策得到一定贯彻执行，也取得一定治理效果。黑河流域草原管理政策的执行缓解了农牧民破坏草原植被进行耕种、放牧的行为，草原载畜量逐步减少，管理政策的实施特别是退牧还草政策的实施使草原生态承载力与环境容量提高，对保护黑河流域草原生态环境有一定积极的作用。

但通过数据获得依然有 34.3% 的受访者认为草原管理政策的整体效果一般，具体从政策对农牧民收入与生态环境改善的角度分析，结论说明黑河流域草原管理政策整体效果并不理想。

从居民收入角度来讲，40.6% 的受访者认为草原政策使农牧民收入仅“有些增加”，甚至 26.6% 的农牧民认为草原政策对收入“没有影响”。因此，调查可以得出，黑河流域当地农牧民近年来收入仅有小幅度的增加，甚至出现草原管理政策对农牧民收入没有影响或者造成收入减少的情况，农牧民对近年来收入增长状况普遍不满意。从草原生态环境改善的角度分析，61.4% 的农牧民认为与 10 年前相比草原生态环境有些改善，草原载畜量的减少及退牧还草政策等的实施，使环境恶化得到一定控制，但是环境改善力度、环境容量的扩大程度均很有限。这两个角度说明当地草原保护管理政策对黑河流域农牧民的收入以及流域生态环境改善的影响效果并不够突出，政策实施还有很长的路要走，需要改变思想重新提出更有效的政策解决黑河流域草原生态环境问题（表 6-4）。

表6-4　黑河流域草原管理政策执行效果

政策效果	选项	频数	相对频率/%
整体效果	非常好	151	14.8
	比较好	381	37.5
	一般	349	34.3
	较差	26	2.6
	非常差	9	0.9
对收入的影响	大幅增加	43	4.2
	有些增加	413	40.6
	没有影响	271	26.6
	有些降低	162	15.9
	大幅降低	22	2.2
对生态环境的影响	大幅改善	91	8.9
	有些改善	624	61.4
	没有变化	152	14.9
	有些降低	36	3.5
	大幅降低	8	0.8

3. 草原管理政策效果原因分析

虽然黑河流域草原管理政策得到一定贯彻执行，但依然存在很多问题，造成政策执行效果不突出，原因包括自然因素与人为因素两个方面。

一方面，自然环境变化的影响。通过表6-3的数据及前文分析可知，气温升高、降水减少是造成黑河流域草原管理政策效果不佳的主要自然原因，同时水资源减少、土地减少、旱灾增加、暴雨减少、冻灾增加等，也是可能影响黑河流域草地生态环境变化的自然因素。虽然自然因素具有人为不可控性，但是气候、自然环境的变化对黑河流域脆弱的生态环境必然产生一定的影响。

另一方面，人为因素的影响，包括农牧民因素、管理因素。农牧民本身具有追求收入、增加放牧与种植力度、破坏草原生态的倾向，同时草原管理者存在管理松懈、处罚力度不够的问题，加剧了农牧民违反草原管理规定、进行草原破坏的行为。综合来说，有以下几个矛盾与问题影响草原管理政策效果。

1）农牧民收入来源单一

通过图 6-3、表 6-5 的数据可以获得，黑河流域农牧民收入主要集中在畜牧业、种植业及外出打工收入，家庭分别占比 22.9%、22.4% 及 9.6%。干旱、缺水的气候与山地为主的地形，造成当地不适宜发展大规模种植业，农业收入有限。同时，外出打工面临很高的机会成本，并且对家庭生活的影响极大，大部分人不情愿外出打工。而草原为主的自然环境使当地适宜发展畜牧业，畜牧业成为当地最主要的收入来源，构成农牧民的支柱性基础收入，因此黑河流域农牧民收入来源单一且集中。如果政府强制实行退牧政策，禁止农牧民放牧，那么农牧民的收入必然受到根本性影响；即使生态环境得到一定改善，农牧民收入的问题也难以解决。给予农牧民一定补助的方式也无法从根源上解决农牧民的收入问题，农牧民依然无法找到新的增加收入的方式，只能再次通过回归放牧，继续破坏草原、环境的方式获得持续性收入，并且由于放养的方式相对于家庭饲养的方式成本低，农牧民更倾向于草原生态破坏更为严重的放养方式。

表6-5　黑河流域农牧民家庭收入主要来源

选项	频数	相对频率 /%
畜牧业	233	22.9
种植业	228	22.4
养殖业	29	2.9
生态补偿	13	1.3
经商	4	0.4
外出打工	98	9.6

2）农牧民找不到其他增加收入来源的方式

由表 6-6 可知，农牧民提出的增加收入来源的方式包括发展副业、外出打工、增加放牧，家庭分别占比 41.1%、19.5% 及 11.3%。在政府禁止放牧的政策下，恶劣的环境不适宜发展农业，外出打工面临一系列的家庭问题，因此相比于农业与打工，农民可以通过养殖家禽、家庭手工业等方式发展副业，这是最现实的增加收入的方式。其次是外出打工的方式。农村存在大量剩余劳动力，由于无法获得其他的收入方式只能通过打工增加收入。除此之外，如果农牧民找不到合适的增加收入的方式，就会采取违反政府政策甚至违法的方式，通过增加放牧、草原破坏来增加收入，这是农

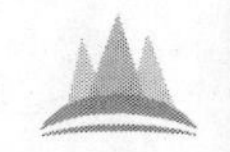

牧民为增加收入采取的极端行为。因为政府找不到增加农牧民收入的其他合理方式，草原管理者往往对农牧民回归放牧、破坏草原采取不管不问、尽量不予惩罚的态度。

表6-6　黑河流域农牧民增加收入的主要途径

选项	频数	相对频率 /%
增加放牧	115	11.3
外出打工	198	19.5
发展副业	418	41.1
移民到外地	5	0.5
不改变	26	2.6

3）政府的监管力度、处罚力度有限

虽然前文提出黑河流域相关草原管理政策得到执行，但执行力度、监督管理、惩罚惩处执行力等均无法保障。对于部分农牧民违反草原管理政策的行为，28.5% 的农牧民从来没有受到处罚，25.6% 的农牧民偶尔受到处罚，说明政府对草原破坏行为的处罚力度极其有限，监管力度一般。对违反草原管理政策的主要原因，农牧民认为是追求收益（26.2%）、管理松懈（22.6%）及处罚力度不够（22.0%）。农牧民收入有限，收入来源单一，农牧民存在增加收入、追求收益的强烈倾向；因为政府无法为农牧民提供切实可行、有效的增加收入的方式，农牧民只能通过最便捷、可行的增加放牧、破坏草原的行为增加收入。鉴于政府无法提出增加农牧民收入的方式，管理者只能对农牧民增加放牧的草原破坏行为睁一只眼、闭一只眼，造成管理松懈、处罚力度不够的状况，这更加剧了农牧民违反草原保护政策、追求个人收益的破坏行为（表 6-7）。

表6-7　农牧民违反草原管理行为状况

题目	选项	频数	相对频率 /%
有没有因为违反草原政策受到处罚	经常	15	1.5
	偶尔	260	25.6
	一般	131	12.9
	有	214	21.0
	从来没有	290	28.5

续表

题目	选项	频数	相对频率 /%
违反草原管理政策的主要原因	追求收益	266	26.2
	管理松懈	230	22.6
	处罚力度不够	224	22.0
	跟随别人	61	6.0
	其他	42	4.2

总的来说，黑河流域草原管理的根本问题是农牧民收入单一的问题。农牧民收入来源单一，难以获得其他增加收入来源的方式，农牧民在综合考虑环境状况、家庭劳动力、收入状况等因素的前提下，倾向于选择继续扩大放牧的方式增加收入，加之政府的监管、惩处力度不够，这造成黑河流域草原环境的持续破坏，形成“收入来源少→扩大放牧→草原破坏→环境恶化→收入降低”的恶性循环，并且很难通过有效的方式打破这一循环（图 6-7）。

因此，黑河流域草原治理面临当地农牧民收入来源单一与草原生态环境保护的矛盾，解决这一矛盾成为黑河流域草原管理政策的核心与关键点。鉴于此，黑河流域草原生态环境治理政策需要建立以增加当地农牧民收入来源作为根本保障的草原保护基础措施，解决当地农牧民收入问题，在解决关键、核心问题的基础上，结合综合的草原保护配套政策机制，才能最终实现黑河流域草原环境的根本治理与改善，达到标本兼治的效果。

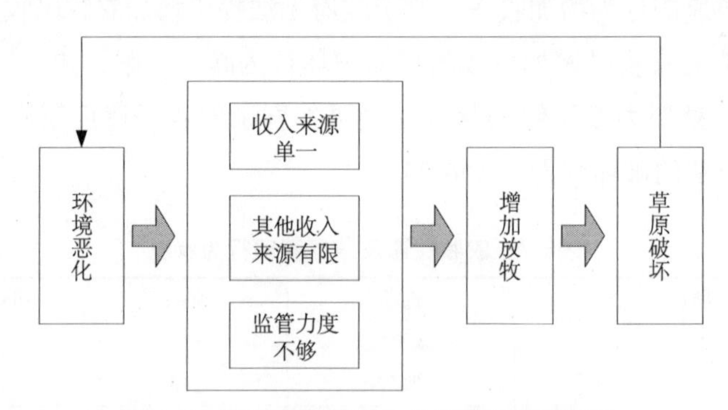

图 6-7　黑河流域草原生态问题恶性循环圈

4. 不同受访者调查结果的差异

草原管理政策对不同特性的农牧民产生不同的政策影响，而不同受访

农牧民的目标不同，对草原管理政策的态度也不尽相同，因而影响草原管理政策的实施效果。因此，受访者基本特征对研究结果会产生不同的影响，草原管理政策的实施也应该综合考虑流域内农牧民的不同状况，通过区别对待使草原管理政策的实施更具有针对性。本书选取受访者年龄、家庭收入两个因素，探索分析草原管理政策对农牧民家庭收入的影响、农牧民违反草原管理政策的原因、农牧民提高收入可行的方式、农牧民认可的草原管理政策的重点四个问题在不同农牧民特征状况下的差异。

1）受访农牧民年龄对分析结果的影响

研究按照年龄特征将受访者分为青年、中年、老年三个等级：年龄低于30岁的为青年，年龄位于30～50岁的为中年，年龄大于50岁的为老年。农牧民年龄特征对研究结果具有一定的影响。首先，根据图6-8的数据，黑河流域草原管理政策对不同年龄段的农牧民家庭收入影响认知有不同的结果，中老年受访者更多地认为政策对家庭收入增加有一定的影响，但年轻受访者却更多地认为政策对家庭收入没有影响。这说明年龄差距对草原管理政策效果的认知不同，中老年对草原管理政策收入效果的认可度更高，认为国家的草原管理政策能够在一定程度上增加农牧民收入，青年人则不认同。其次，根据图6-9的数据，不同年龄段的农牧民对部分农牧民违反草原管理政策的原因认同也有显著的差异，青年人、中年人更多地认为是由于农牧民追求收益的原因，而老年人则主要认为是政府处罚力度不够的问题。而根据图6-10、图6-11的数据，不同年龄阶段在提高未来收入的方式、未来草原管理政策的重点方面的观点没有显著的差异，说明年龄对两方面的观点没有差异。因此，综合来说，年龄差异在农牧民对草原管理政策评价方面产生一定的影响，农牧民年龄越小，认为草原管理政策对增加收入的影响越小，并且越更多地认为破坏草原的行为是收入问题引起的，而不是管理的问题，也说明年轻人对收入增加有更大的期望，对草原管理政策存在更大的反对与质疑。

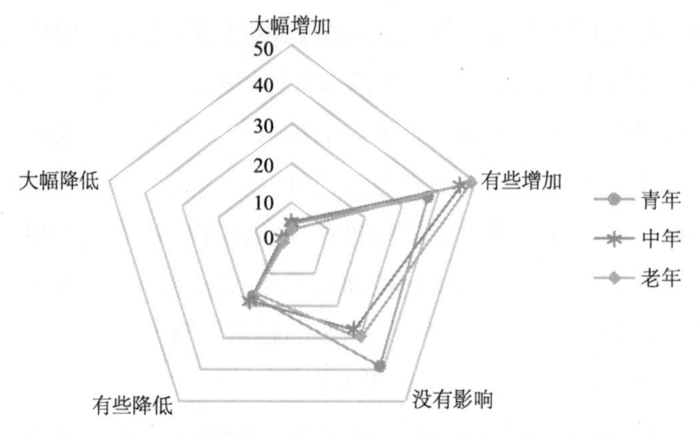

图 6-8　不同年龄段农牧民草原管理政策对收入影响的认知差异

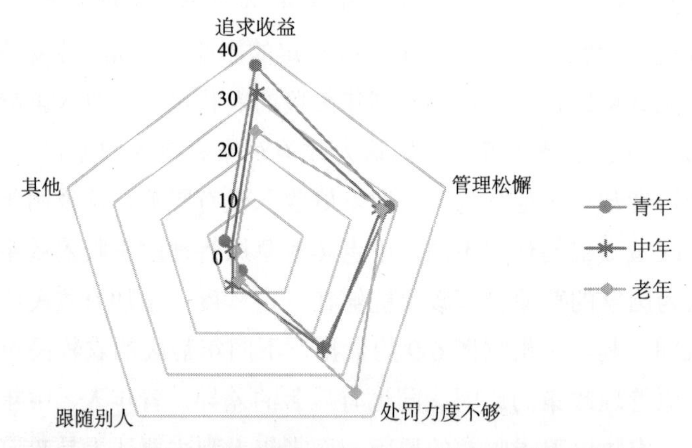

图 6-9　不同年龄段农牧民对违背草原管理政策原因的认知

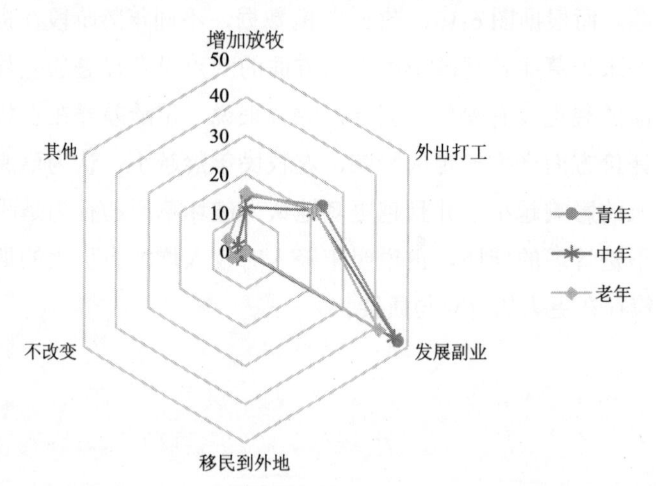

图 6-10　不同年龄段农牧民增加收入方式的差异

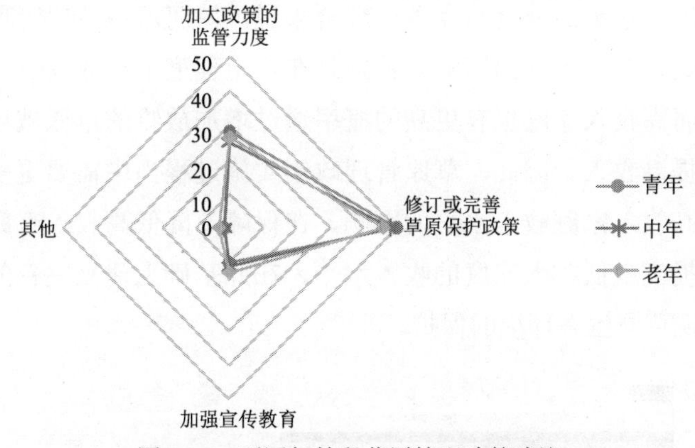

图 6-11　不同年龄段草原管理政策建议

2）受访农牧民家庭收入水平对分析结果的影响

研究将农牧民按照家庭年均纯收入分为三类：家庭年纯收入低于 3 万元的为低收入家庭，年纯收入为 3 万～ 7 万元的为中等收入家庭，年纯收入大于 7 万元的为高收入家庭。家庭收入对研究分析结果也具有一定的影响。首先，通过图 6-12 的数据结果可得：不同收入阶层的家庭在草原管理政策对家庭收入的影响方面有不同的看法，收入水平越低的家庭越多地认为草原政策能够使家庭收入有些增加，而收入水平越高的家庭，草原政策对家庭收入的影响越小，甚至有较高比例的家庭由于草原保护政策的实施家庭收入有所降低，因此草原保护政策对低收入家庭收入增加有一定的正向影响，而对高收入家庭收入的增加具有一定的负面影响。其次，通过图 6-13 的数据，虽然不同收入阶层的家庭均认同农牧民违反草原管理政策主要由于追求收益、管理松懈与处罚力度不够三个方面的问题，但是收入水平越高的家庭更多地认为追求高收入是农牧民违反草原管理政策、破坏草原环境的原因，而中低收入家庭则不够明显。再次，通过图 6-14 的数据获得，不同家庭均主要希望通过发展副业增加收入，但低收入家庭对发展副业有更大的渴望，而有相当比例的高收入家庭希望通过增加放牧的草原破坏行为增加收入。这说明高收入家庭对草原保护具有更高的破坏性，对草原保护政策具有更高的违背概率。最后，通过图 6-15 的数据，不同的家庭均认为未来草原管理的重点是通过在修订和完善草原保护政策的基础上加强监督管理的方式实现草原保护，说明农牧民对草原保护政策普遍存在质疑，需要进行政策的修订与完善。综合来说，黑河流域草原管理政策对不同收入家庭均产生不同的影响，特别是在一定程度上造成高收入家庭的收入降低，而高收入家庭也有更高的概率通过增加放牧的草原破坏与政策违背行为来提高收入。因此，草原管理政策实施过程当中需要更多地重视政策对一定高收入家庭收入水平的影响，在保障不降低高收入家庭收入水平的基础上提高中低收入家庭的收入水平，在此基础上建立完备的草原保护政策并实现草原生态环境的保护。

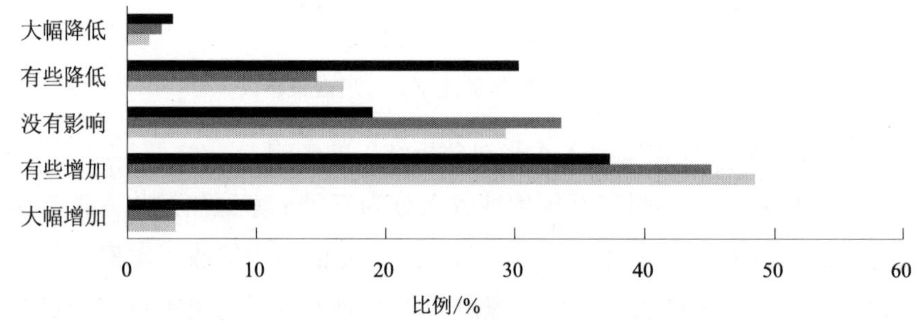

	大幅增加	有些增加	没有影响	有些降低	大幅降低
■高收入	9.9	37.3	19.0	30.3	3.5
■中等收入	3.7	45.1	33.6	14.8	2.6
■低收入	3.7	48.5	29.3	16.8	1.7

图 6-12　不同收入阶层农牧民草原管理政策对收入影响的认知差异

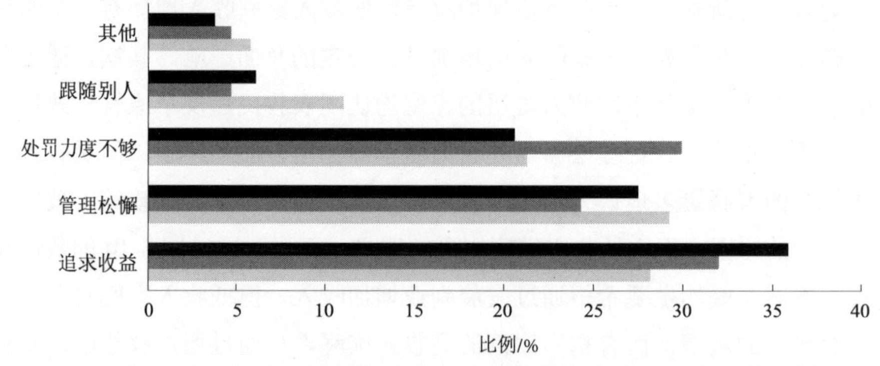

	追求收益	管理松懈	处罚力度不够	跟随别人	其他
■ 高收入	35.9	27.5	20.6	6.1	3.8
■ 中等收入	32.0	24.3	29.9	4.7	4.7
■ 低收入	28.2	29.2	21.3	11	5.8

图 6-13　不同收入阶层农牧民对违背草原管理政策原因的认知

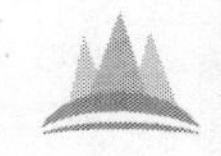

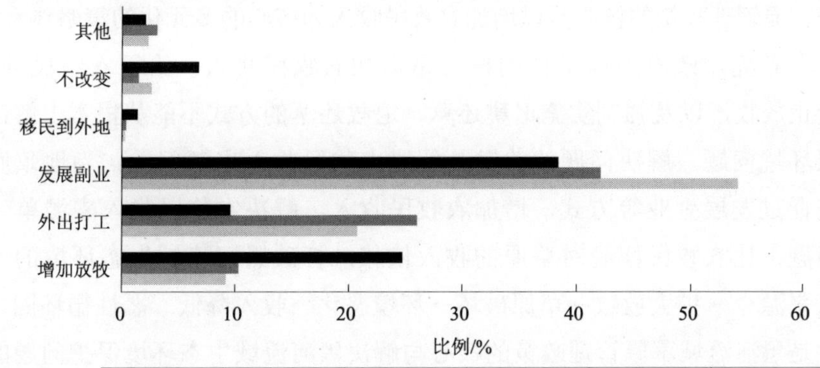

	增加放牧	外出打工	发展副业	移民到外地	不改变	其他
■ 高收入	24.7	9.6	38.4	1.4	6.8	2.1
■ 中等收入	10.3	26.0	42.1	0.4	1.5	3.1
■ 低收入	9.2	20.7	54.1	0.3	2.6	2.3

图 6-14　不同收入阶层农牧民增加收入方式的差异

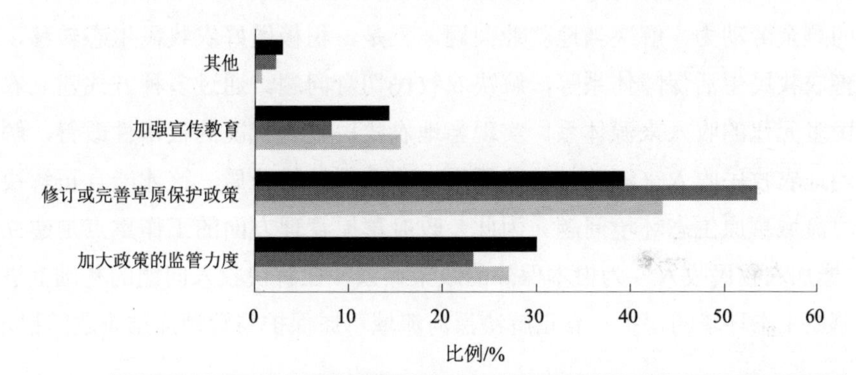

	加大政策的监管力度	修订或完善草原保护政策	加强宣传教育	其他
■ 高收入	30.0	39.3	14.3	2.9
■ 中等收入	23.3	53.4	8.1	2.2
■ 低收入	27.1	43.3	15.5	0.7

图 6-15　不同收入阶层草原管理政策建议

6.3　草原管理重点及政策建议

6.3.1　政府草原管理政策工作核心

黑河流域草原管理存在的根本问题是农牧民收入单一的问题，因此未来草原管理政策的核心是以增加农牧民收入为核心的多元化的策略体系。

首先，政府当前工作的核心是增加农牧民收入。给予农牧民补助、禁止放牧，以及强制实施退耕还草、退牧还草的方式不能从根本上解决草原环境问题，解决问题的关键是解决农牧民收入来源问题。当地政府应该通过发展副业等方式，增加农牧民收入，解决农牧民收入来源单一的问题，让农牧民摆脱对草原的收入依赖，打破黑河流域生态环境的“收入来源少→扩大放牧→草原破坏→环境恶化→收入降低”恶性循环圈。这才是黑河流域草原管理政策的核心与解决黑河流域生态环境保护问题的根本。其次，政府未来工作的重点是建立农牧民多元化的收入来源体系。一是，政府应该通过政策引导帮助农牧民发展副业，建立当地的第三产业体系，拓展当地农牧民收入来源，摆脱当地农牧民对畜牧业、对草原生态的依赖。二是，政府通过引入制造业等新的产业增加就业，吸纳因禁牧释放的剩余劳动力，解决当地就业问题。三是，积极做好农牧民生态转移，完善农牧民生活保障体系等，解决农牧民切身问题。通过多种方式建立农牧民多元化的收入来源体系，实现当地农牧民收入状况的根本性改善，解决当地农牧民收入来源单一与草原生态环境保护的矛盾，这才能真正解决黑河流域草原生态环境问题。因此，政府草原管理方面的工作重点是建立以“增加农牧民收入”为根本保障的政策体系，在解决收入问题的基础上进行草原生态环境的保护，真正解决黑河流域草原保护与管理症结难题的根本。

6.3.2 黑河流域草原管理政策建议

1. 黑河流域草原管理政策的基础是在修订和完善草原保护政策的基础上加强监督管理

对未来政府草原管理的工作重点的调查，42.6%的农牧民认为首先应该“修订和完善草原保护政策”，在政策完善的基础上加大草原保护的监管力度和宣传教育（分别占比22.6%和10%）。黑河流域草原管理存在的问题在于政策本身存在的问题，以往草原管理进行的直接补贴、禁止放牧，以及强制退牧还草、退耕还草的措施对于草原生态保护只能治标不治本，并不是农牧民本身缺乏草原保护意识，破坏草原、增加放牧是农牧民为了增

加收入采取的极端行为。政府应该在修订和完善草原管理与保护政策的基础上采取其他措施，这才是解决黑河流域草原管理困境的基础（表6-8）。

表6-8　黑河流域草原管理未来工作重点

选项	频数	相对频率 /%
加大政策的监管力度	230	22.6
修订和完善草原保护政策	425	41.8
加强宣传教育	102	10.0

其次，完善草原保护政策必须加强监督管理。原有政策对草原、林区、牧区保护的执法力度不强，应加强监管巡查，加强法律宣传，加强处罚力度。同时，加强对专项治理建设资金的监管，防止挪用现象，促进政策透明等。在政策完善、监管加强的基础上，促进黑河流域草原管理机制的完善以及草原生态环境的改善。

同时，要做到根据农牧民的不同特征，建立具有针对性的政策体系。前文的分析结果表明，草原管理政策对不同收入家庭、不同年龄农牧民产生不同的影响，而农牧民及家庭不同的状况也会影响草原管理政策的实施以及政策效果的实现。因此，草原管理政策的实施应该根据不同家庭状况、不同个人状况建立具有针对性、适用性的政策体系，保障高、中、低收入家庭生活水平均能够实现明显的改善，不因为草原政策的实施使任何人的利益受到损害。

2. 黑河流域草原管理政策的核心是增加农牧民收入

黑河流域农牧民总体收入水平不高、差距大，需要增加农牧民收入，改善生活条件。政府应该通过政策引导帮助农民发展副业、建立当地的第三产业体系，拓展当地农牧民收入来源；通过引入制造业等新的产业增加就业，吸纳因禁牧释放的剩余劳动力，解决当地就业问题。通过多种方式实现建立农牧民多元化的收入来源体系，最终实现当地农牧民收入状况的根本性改善，解决当地农牧民收入来源单一与草原生态环境保护的矛盾。

产业结构调整是促进农牧民收入增长的主要途径。畜牧业是黑河流域的产业基础，应该建立合理有序的畜牧业发展方式，提高牧民收入。在销售方面，保证有平稳持久的销售市场；在产业链方面，政府对畜产品进行深加工，增加产业附加值；在养殖方式方面，持续稳定地实行舍饲、半舍饲的规范化养殖，同时引进推广多胎肉羊，尽量减少放牧，推进规模化圈养。同时，政府应该帮助牧民发展第三产业，发展副业，推广种植苁蓉等经济作物，并且对农民自主创业给予一定的资金支持等方式。通过多元化的方式增加黑河流域农牧民收入来源。

3. 切实解决农牧民关心关注的重大问题

草原保护政策的实施不仅应该在完善政策体系、增加农民收入、加强监督管理方面做好工作，还应该通过多元化的政策完善黑河流域草原生态区整体的社会发展，解决与农牧民切身相关的重大问题。通过对农牧民的调查分析，应包括以下几个方面。

（1）调整奖励补偿政策。调整奖励补偿制度，提高草原补偿、草畜平衡补助、人工草地等政策的补助补偿标准。同时加大对农牧民的养殖业补助、农机具补助、水资源补偿与用水补助，以及对草原管护人员的补助、少数民族大学生的补助等。通过加大补助标准，扩大补助范围，保障农牧民生活、生产条件。

（2）健全草原管护管理。草原环境逐渐恶化，面临资源减少、草场退化、环境干燥等问题，严重影响农牧民的生产生活，需要加强草原生态保护，健全草原管护制度。草原管护由县乡村统一管理，提高草原补助奖励标准，同时加强虫灾（蝗虫）治理、加强水源保护力度。

（3）基础设施建设。政府应该加大基础设施建设的投资力度，包括：加强对饮用水设施的建设，兴建水库，合理开发地下水资源，解决农牧民饮水问题；保障道路畅通，加强道路建设；进行水电站建设，保证生活用电；加强农村环境卫生整治与农村规划，改善农村环境。

（4）加强社会保障。社会保障关乎农牧民切身利益，需要完善社会保障，保证农牧民基本生活。社会保障具体包括：加强户口与草原政策的结合；草原教育质量差，教育花费大，应该切实解决农牧民子女教育问题；提高医疗水平，解决医疗问题；在就业方面，对农牧民子女就业有一定的倾向支持；在住房方面，尽量解决禁牧退牧人员转移搬迁安置房问题；健全养老机制。

第7章 基于管理者调查问卷的草原管理政策分析

7.1 调查问卷设计

7.1.1 调查内容

研究同时进行针对黑河流域相关管理者为调查对象的“黑河流域草原管理政策执行情况”调查，设计“中国西部草原管理政策研究——管理者政策执行情况调查”调查问卷（附录2管理者调查问卷）。问卷通过了解相关管理者对黑河流域草原管理政策执行的情况，获取真实信息与数据，为本项目的相关研究提供重要的支撑；同时，调查结果将与针对农牧民调查的结果进行比较与分析，获取相关探索结果。

调查问卷的设计以获得黑河流域相关管理者的草原管理政策执行情况为目的，通过研究组成员讨论设计、相关领域专家的建议讨论，获得管理者调查问卷。问卷包括三部分，即被调查人的基本信息、生态环境与政策执行情况、政策执行效果，而每个部分又设计多个题目来获取相关研究的具体状况（具体调查问卷见附录2）。

第一部分为被调查人的基本信息，反映被调查管理者的民族、性别、年龄、行政级别、文化程度、政治面貌、工作年限及所属部门类别等信息。被调查者的基本状况将影响调查结果的有效性与真实性，在调查过程当中应该尽量选择年龄偏长、文化程度较高、工作年限较长的基层管理者，同时保障被调查者在民族、性别、行政级别及所属部门等方面的差异性与区分度，以保障调查结果的真实度与有效性。第二部分为生态环境与社会政策调查情况，反映黑河流域近年来生态环境的变化以及产生变化的自然原因，同时对黑河流域草原管理政策的执行情况进行调查。第三部分为草原管理政策执行情况调查，包括政策执行效果及产生的原因，同时对管理者监管、处罚等情况进行调查。调查结果将用于黑河流域草原管理政策的执行效果分析，并且与农牧民问卷的草原管理政策执行情况结果进行比较探索，为研究提供数据支撑（表 7-1）。

表7-1　管理者调查问卷主要内容

调查方面	调查内容
被调查人基本信息	基本信息：民族、性别、年龄、行政级别、部门类别、文化程度、政治面貌、工作年限 家庭状况：家庭人口、劳动人口
生态环境与社会政策调查	生态环境：气温变化、降水变化、其他因素 草原生态保护政策评价 草原生态保护补偿奖励情况
草原管理政策执行情况	政府在草原保护方面的态度 本地对草原保护政策或措施的监管力度 草原破坏的处罚力度 违反草原管理政策的主要原因 草原保护政策或措施对收入的影响 草原保护政策或措施的整体治理效果 政府未来在草原管理方面的工作重点

7.1.2　调查方案

为保证调查结果的真实性，项目采用实地调查方式。来自中国科学院科技政策与管理科学研究所及兰州大学的本课题组相关研究人员通过深入黑河流域的相关地区，与基层管理者及省市相关单位管理者进行面对面的交流访问，获取与研究相关的一手数据资料，保证调查结果的真实性、有效性、说服力。

调查基本覆盖黑河流域的全部县（区、旗）的管理者，覆盖区域广泛，调查数据真实可靠。调查问卷共发放 250 份，回收 250 份，有效问卷 233 份，其中甘肃省 183 份、内蒙古自治区 42 份、青海省 8 份，问卷有效率达 93.2%。从具体调查单位来看，包括省属、市属相关单位，以及具体各县（区、旗）相关草原管理机构的行政人员。从具体各县（区、旗）、市分布

来说，内蒙古额济纳旗42份，青海省祁连县8份，甘肃省农牧厅2份、金塔县24份、甘州区50份、高台县30份、临泽县18份、民乐县44份、肃南县15份。问卷覆盖地区包括黑河流域全部县（区、旗），并且涵盖兰州市、张掖市等省市相关机构部门，地区分布合理，能够真实反映研究需要探索的问题。

7.2 调查问卷结果

7.2.1 被调查人的基本信息

1. 基本特征

研究问卷的调查对象是黑河流域相关地区部门管理者，大部分从属于畜牧、林业、草原等部门，从事草原监管与执法、畜牧管理与技术推广、行政管理等具体工作。从民族构成来看，以汉族为主，占比76.8%，蒙古族、裕固族也占有一定比例；从性别来看，以男性为主，占比70%以上；从年龄来看，41～50岁管理者占比最高，为39.1%。民族、性别、年龄分布合理有效。从政治面貌来看，60%以上的受访者为中共党员；从行政级别来看，副科及以下占比63.1%，说明受访者以基层中共党员管理者为主，能够深入基层，对农牧业生产、草原管理具有更真实的了解。从文化程度来看，70%以上为大学（大专）以上文化水平，受教育水平较高，对农牧业生产、草原管理现状及存在的问题具有更深入的认识。从工作年限来看，工作10年以上的管理者占50%以上，说明大部分被调查对象的管理者工作经验丰富，对草原状况、草原管理政策了解全面、深刻，能够保证问卷信息的真实性与有效性。通过对被调查人基本特征的统计可以看出，调查对象分布特征合理有效，能够真实、有效地反映本项目将要研究的问题。

2. 家庭状况

对管理者家庭状况进行统计，家庭平均总人口3.5人，劳动人口2.2人，其中从事农业生产1.88人、从事牧业生产0.73人。家庭平均从事农业生产的人口远高于家庭平均从事牧业生产的人口，说明当地基层管理者的家庭劳动者更倾向于从事农业生产，从事牧业生产的基层管理者家庭劳动力较少（图 7-1）。

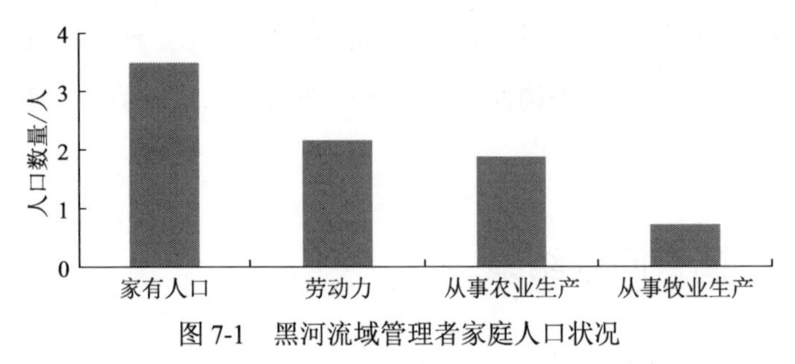

图 7-1　黑河流域管理者家庭人口状况

7.2.2　草原管理政策评价

1. 草原管理政策执行情况

草原管理政策包括退牧还草政策，退耕还林政策，禁牧、休牧、轮牧政策，草畜平衡政策，奖励补偿政策等五大类；而生态补偿政策包括禁牧补助、退牧补助、退耕补助、草畜平衡奖励、人工草地补助、畜牧良种补贴、牧草良种补贴、生产资料综合补贴。根据管理者建议综合五个政策的重要性排名大致为，奖励补偿政策＞禁牧、休牧、轮牧政策＞退耕还林政策＞退牧还草政策＞草畜平衡政策。各项政策实施较为合理，并且得到较好的贯彻执行，特别是相对重要的草原生态保护补助及奖励政策得到很好的贯彻执行。因此，从管理者的角度来说，政府一直比较重视草原生态保护，各个政策得到贯彻执行，特别是政府在草原生态保护奖励补偿政策方面执行比较完善，建立起完善的草原生态保护补偿奖励机制。

2. 草原管理政策执行效果

调查受访管理者关于草原保护政策的治理效果，46.4% 的受访者认为政策整体治理效果比较好，但依然有 24.9% 的管理者认为草原保护政策的治理效果一般。具体从政策对当地农牧民收入、草原生态变化的影响来看：一是政策对农牧民收入的影响，最多的受访管理者认为农牧民收入有一定增加（36.1%），有 16.3% 的管理者认为草原政策对牧民收入增加并没有什么影响，甚至有 18.9% 的受访管理者认为草原管理政策使牧民收入有些减

少。二是政策对当地草原生态环境变化的影响，通过调查统计，40.8% 的受访管理者认为10年来草地状况有一点改善，但是也有一定比例的受访者（13.7%）认为草原环境有一点恶化。同时，生态环境恶化主要表现在四个方面：草原面积减少、草地种类减少、劣质牧草增加和土壤恶化（表7-2）。

表7-2　管理者草原政策执行效果

政策效果	选择	频数	相对频率 /%
总体治理效果	非常好	16	6.9
	比较好	108	46.4
	一般	58	24.9
	较差	4	1.7
	非常差	1	0.4
收入的影响	大幅增加	18	7.7
	有些增加	84	36.1
	没有影响	38	16.3
	有些减少	44	18.9
	大幅减少	5	2.1
生态环境的变化	有很大改善	36	15.5
	有一点改善	95	40.8
	基本没变	12	5.2
	有一点恶化	32	13.7
	有很大恶化	19	8.2
生态环境变化的表现	面积变小	40	17.2
	种类减少	32	13.7
	劣质牧草增加	15	6.4
	土壤恶化	30	12.9

通过以上针对管理者对黑河流域草原管理政策执行效果的调查结果可以获得结论：在草原管理政策的综合效果、政策对环境改善的效果、政策对农牧民收入的影响三个角度，政府草原保护与管理政策虽然取得一定效果，但依然存在很多不足，环境改善有限、政策对农牧民收入的增加有限，造成草原环境政策综合治理效果有待提高。

3. 政策效果原因分析

黑河流域草原管理政策执行效果不足，原因包括自然因素与人为因素两个方面。

影响黑河流域草地生态环境恶化的自然原因包括降水减少、气温升高、水资源减少、土地减少、旱灾增加、暴雨减少、冻灾增加等因素。其中，气温、降水是影响草原生态最主要的自然因素，通过统计，54.5% 的受访者认为降水减少以及 51.5% 的受访者认为气温增加是草原环境变化的原因；同时，其他自然因素也直接或间接影响黑河流域草地的状况，成为当地草地生态变化的影响因素。

根据调查统计，人为因素需要从管理者与农牧民两个角度分析。在管理者的角度，黑河流域草原管理过程当中存在处罚力度不够（22.3%）、管理松懈（12.0%）的问题；而在农牧民的角度，农牧民具有追求收益、增加收入从而增加放牧、草原破坏的倾向与动力（18.9%），因而两种共同造成了黑河流域草原生态破坏的人为因素。

表7-3　管理者黑河流域草原政策效果原因分析

人为因素	选项	频数	相对频率 /%
农牧民违反政策的原因	提高收益	44	18.9
	管理松懈	28	12.0
	处罚力度不够	52	22.3
	跟随别人	11	4.7
	其他	27	11.6
监管力度	很强	25	10.7
	较强	91	39.1
	一般	59	25.3
	较弱	12	5.2
	很弱	4	1.7
处罚力度	很多	10	4.3
	有一些	99	42.5
	比较少	61	26.2
	几乎没有	21	9.0

统计结果表明，对于监管力度，39.1% 的管理者认为监管力度较强，但也有 25.3% 的管理者认为政策措施的监管力度一般，仅有 10.7% 的管理者认为政策监管力度很强，可以看出，当地政府对于草原管理与保护政策实施的监管力度一般。同时，对于农牧民违反草原管理政策的现象，仅仅有一些会受到处罚（42.5%），大部分草原政策违反行为并没有受到处罚，因此当地政府对农牧民草原破坏行为的处罚力度很有限。综合分析可以看出，虽然当地政府对草原保护的态度比较重视，但是依然存在农牧民违反

草原保护条例、破坏草原的情况，造成政策的治理效果一般。造成这种状况最主要的原因在于处罚力度不够，应该通过加强草原管理、加大处罚力度杜绝农牧民草原破坏的行为。其次，针对农牧民收入不足的状况，应该努力通过一系列措施增加农牧民收入，杜绝农牧民无节制地扩大畜牧业生产、破坏草原的行为（表7-3）。

7.2.3 政府草原管理的工作重点

从管理者角度分析未来草原管理的工作重点，包括修订和完善草原管理与保护政策、加大现有政策的实施力度、提高牧民收入三个方面。

首先，修订和完善草原管理与保护政策。现有的草原管理政策存在一定的问题，没有抓住草原管理问题的核心，造成草原管理效果较差。因此，未来政府工作的首要任务是修订和完善草原管理与保护政策，建立以增加农牧民收入为核心目标与基础的新的政策体系，在此基础上加大政策实施力度、加强农牧民草原保护意识宣传，最终提高政策实施的整治效果。

其次，加大现有政策的实施力度，加强监管。政策实施不够完善、处罚力度不足、监督管理不足等，加剧了农牧民的草原破坏行为。强化草原监管具体包括：加强草原监管队伍与人员编制建设，加强资金、人员、配置的投入，健全草原监管机构，通过加强管理投入，实现草原监督管理水平的提高。强化政策实施具体包括：保障政策实施的连续性，增加草原奖励与补偿资金，提高基层工作人员素质，加强草原保护政策的宣传力度，加大草原破坏行为的惩处力度等。同时，切实保障草原管理人员的利益，解决基层管理人员的保障、待遇等问题。

再次，努力提高农牧民收入。并不是农牧民草原保护意识不够，而是农牧民收入有限并且难以获得增加收入的其他途径，往往采取增加放牧、破坏草原的行为增加收入。因此，未来草原管理政策的核心应该以农牧民的利益为突破点，努力增加农牧民收入，在农牧民收入增加、生活改善的基础上改善草原生态环境。同时，保障农牧民生活需要解决农牧民饮水、住房、教育、医疗等基本的社会保障问题，转变农牧民生产、生活方式，提倡舍饲养殖，减少放牧等。

最后，在增加农牧民收入、改善农牧民生活的基础上，加强环境治理。做到：遏制环境污染问题，降低农业污染；减少耕地、化肥对环境、土壤造成的影响；治理病虫害问题等。通过自然环境治理角度、人为角度的环境治理策略体系，实现黑河流域环境的综合治理与改善。

7.3 农牧民与管理者问卷结果比较

7.3.1 调查结果比较

1. 草原管理政策效果比较

1）草原管理政策执行情况比较

通过对比农牧民问卷与管理者问卷对黑河流域草原管理政策的执行情况，可以获得大致相同的结论。我国西部草原管理政策包括退牧还草政策，退耕还林政策，禁牧、休牧、轮牧政策，草畜平衡政策，奖励补偿政策等五大类，各个政策均得到相对完善的贯彻实施，说明政府在草原管理政策实施方面做出了一定的努力与工作。特别是，草原生态保护补助及奖励政策机制得到了较好的贯彻执行。因此，综合来说，我国在西部地区的草原管理政策特别是国家对奖励补偿、财政支出方面的政策得到较好的贯彻执行。

2）草原管理政策执行效果比较

通过两份问卷的比较可以获得相同的结论。总的来说，两份问卷均说明政府的草原管理政策取得一定的管理效果，特别是草原管理政策的执行一定程度上遏制了农牧民破坏森林、草原进行耕种、放牧的行为；通过管理政策的实施，特别是退牧还草政策的实施草原载畜量减少，对保护黑河流域草原生态环境有重要的作用。黑河流域草原保护政策取得的政策效果是包括当地农牧民及基层管理者在内的相关人员有目共睹的事实；也说明国家致力于我国西部地区生态保护、生态恢复与环境改善，进行草原、森林、绿地等生态环境保护的资金、人力等投入是有意义的。

但是，两份问卷也均表明草原管理政策存在一定的问题，整体治理效果欠佳。从整体治理效果来看，有较多的受访者认为草原管理政策的治理

效果一般，而具体分析政策对生态环境的改善、对农牧民收入的增加状况，政策也没有取得预期的效果。针对政策对农牧民收入的影响，有较多的受访者认为草原政策对农牧民收入没有产生影响，甚至有一定的受访者认为政策造成农牧民的收入有所降低。同样地，有一定数目的受访者认为草原管理政策对黑河流域草原生活环境的变化没有影响，甚至在一定程度上有所变差。因此，综合来说，黑河流域草原管理政策取得一定的效果，特别是对遏制环境恶化的态势具有重要意义，但是对草原生态环境改善、农牧民生活改善效果不突出，这说明我国西部地区草原管理政策本身存在一定的问题，需要继续完善政策，也说明我国对西部地区生态环境的保护方面在未来还需要很长的道路要走，任重而道远（图 7-2 ～图 7-4）。

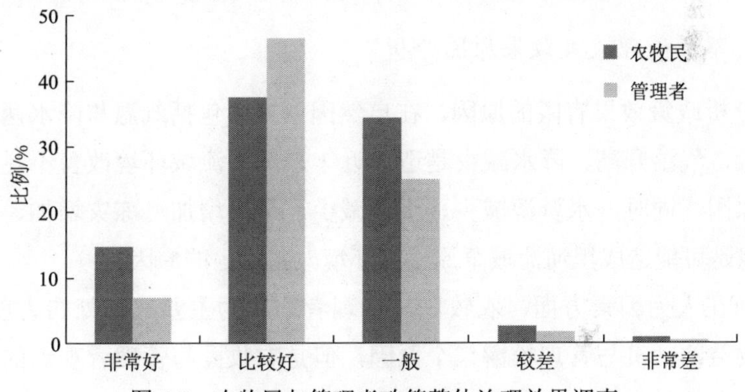

图 7-2　农牧民与管理者政策整体治理效果调查

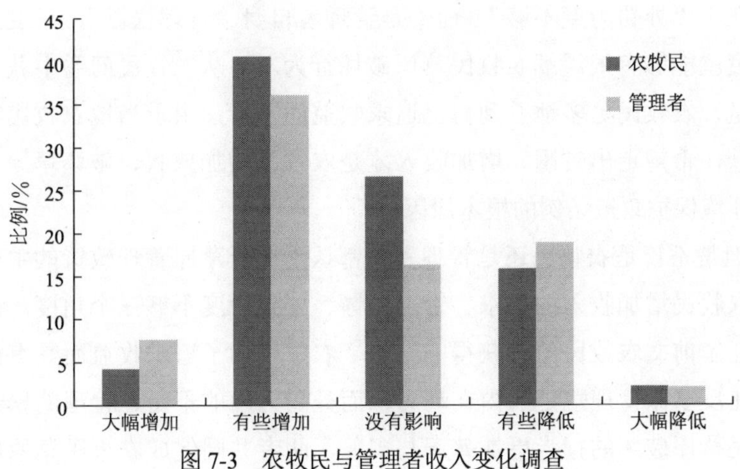

图 7-3　农牧民与管理者收入变化调查

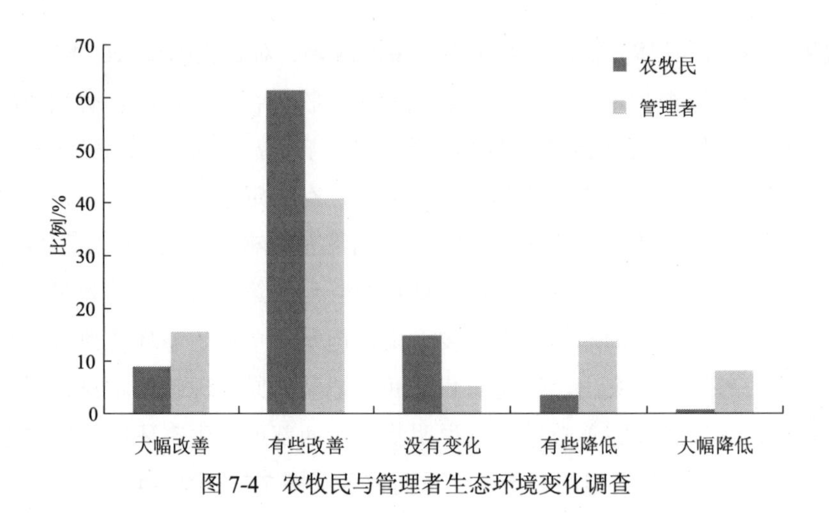

图 7-4 农牧民与管理者生态环境变化调查

2. 草原管理政策效果原因分析

分析政策效果有限的原因，在自然因素方面包括气温与降水两个主要的因素，气温升高、降水减少是造成近年来黑河流域环境改善不足的主要自然原因。同时，水资源减少、土地减少、旱灾增加、冻灾增加、暴雨减少等也是可能造成黑河流域草原生态环境改善不足的原因。

而在人为因素方面，农牧民与管理者均认为主要包括处罚力度不够、提高收益的动机与管理松懈三个原因，但是农牧民与管理者观点的侧重点存在一定的差别。从管理者的角度，认为农牧民违反草原管理政策的主要原因在于“处罚力度不够”，而不是管理者自身“管理松懈”，只要加大处罚力度就能够有效缓解农牧民草原破坏行为。而从农牧民问卷中获得的结论却是，农牧民更多地看到自己追求收益的目标，由于当地农牧民收入来源单一，希望走出贫困、增加收入才是农牧民增加放牧、破坏草原、违反草原环境保护政策条例的根本原因。

但是不管是农牧民还是管理者，都认为违反草原管理政策的主要原因包括农牧民增加收入的追求、管理松懈、处罚力度不够三个角度。两份问卷验证了前文农牧民问卷获得的结论：农牧民为了追求收益而具有破坏草原、违反草原管理规定的内在动力，而处罚力度的不足、管理的松懈造成农牧民草原破坏的行为更加普遍与猖獗，两者共同促成农牧民草原破坏的行为，最终造成黑河流域草原管理政策的实施效果一般。

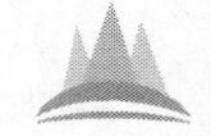

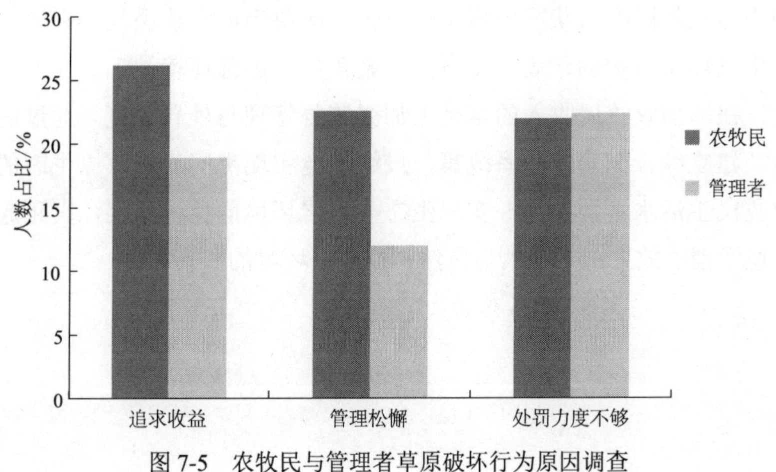

图 7-5　农牧民与管理者草原破坏行为原因调查

7.3.2　基于比较的结论

基于前文农牧民、管理者关于黑河流域草原管理政策效果不足的原因分析，得到相同的问题与结论，即农牧民追求收益、管理松懈、处罚力度不够三个因素共同促成农牧民草原破坏的行为以及草原管理政策效果的不足。其中，最核心、最根本的问题是草原管理政策实施忽视当地农牧民自身收入来源问题，造成农牧民过度放牧、草原破坏行为频现。政府的草原保护措施重视草原生态管理与保护，忽视草原居民的收入与生活问题；仅仅通过单纯的奖励补偿政策、简单粗暴的惩罚措施等只能解决表面的、一时的问题，不能从根源上解决黑河流域草原生态治理的问题，也无法做到长久的生态保护。已有的政策只能起到治标不治本的效果，往往造成政策效果有限，并已形成一种“收入来源少→扩大放牧→草原破坏→环境恶化→收入降低”的恶性循环状态。解决问题的根本出路与出发点应该在于通过一系列措施首先解决当地农牧民收入来源单一的问题，增加农牧民收入，在提高当地农牧民收入的基础上，配合以综合的措施，才能实现黑河流域草原生态环境的综合治理，起到标本兼治的效果。

因此，黑河流域生态保护问题的解决应该通过引导和帮助农民发展副业、建立当地的第三产业体系、引入制造业等新的产业，建立农牧民多元化的收入来源体系，解决当地就业问题，拓展当地农牧民收入来源，最终实现当地农牧民收入状况的根本性改善，解决当地农牧民收入来源问题与草原生态环境保护的矛盾。在自然因素角度，加强环境治理；在人为因素角度，在增加农牧民收入的基础上加强监督管理与处罚力度。通过两者的结合，建立综合配套的体系政策，最终才能实现黑河流域草原生态的改善与农牧民生活水平的提高，实现生态与人居环境的良性发展，真正起到生态环境治理的效果，实现人与自然、资源、环境的和谐相处。

第8章

西部草原可持续发展管理问题及对策

8.1　西部草原可持续发展面临的问题

8.1.1　草原保护与牧民收入

自从国家实施生态保护相关政策以来，草地资源的开发利用受限，相对应的牧业发展受控，在一定程度上影响了牧民收入，制约了草原牧区经济社会的发展。根据本书对黑河流域草原农牧民的调查，黑河流域农牧民家庭收入主要来源于畜牧业、种植业及外出打工三个方面，其中家庭平均畜牧业收入为 33 310 元、种植业收入为 24 968 元、外出打工收入为 20 686 元。对于家庭年均纯收入，3/4 的家庭年均纯收入为 1 万～ 7 万元，其中 3 万～ 5 万元的家庭占比最高，达到 30% 以上，大部分家庭属于中低收入家庭。可以从调查结果上看出，畜牧业收入在家庭收入中占主体，因此禁牧、休牧等草原保护政策对于牧民收入的影响是显而易见的。

牧民在面对如此高昂的机会成本面前，保护草原生态环境的积极性会大大降低，在政策面前，“偷牧”“夜牧”的现象屡禁不止。根据《2014 年全国草原违法案件统计分析报告》统计，2014 年，全国各类草原违法案件发案 18 998 起，立案 17 848 起，立案率为 93.9%；结案 17423 起，结案率为 97.6%。非法开垦草原、非法征收征用使用草原和非法临时占用草原三类案件共破坏草原面积 31.37 万亩。买卖或非法流转草原面积 1.83 万亩。

移送司法机关处理的案件621起，被提起行政复议或行政诉讼的案件25起。从各类草原违法案件的发案数量看，违反禁牧休牧规定案件发案数量14 899起，占发案总数的78.4%，位居第一；非法开垦草原案件发案数量1910起，占发案总数的10.1%，位居第二；违反草畜平衡规定案件发案数量1013起，占发案总数的5.3%，位居第三（图8-1）。

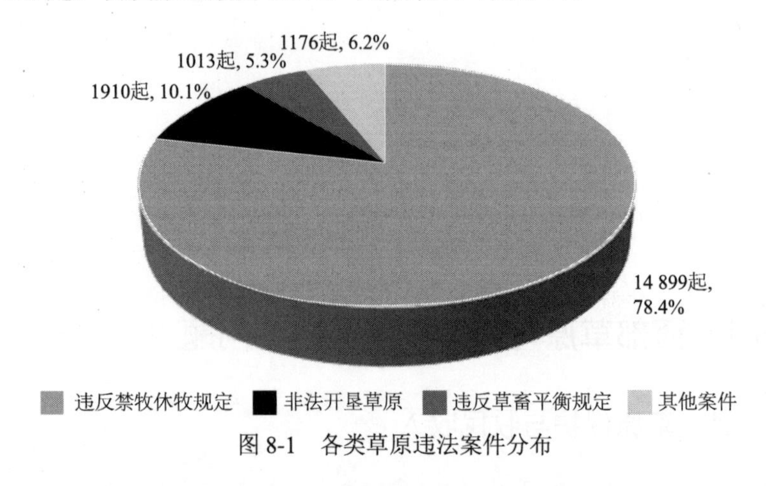

图8-1　各类草原违法案件分布

由此可见，提高农牧民收入是草原管理面临的主要问题。通过拓宽农牧民增收渠道，应增强自主保护意识，实现草原生态持续改善和牧区经济健康发展。

8.1.2　保护政策的短期性、不稳定性特征与牧民长期生计

近年来，草原生态环境问题备受社会各界的重视。人们在寻找草原生态退化根源的同时，也分析草原退化对当地牧民生计产生的影响。有研究认为，滥垦草原、盲目移民的结果使草原生态环境遭到严重破坏而沙化退化（包玉山和周瑞，2001）。有的则认为政策导向的失误、掠夺性开发和粗放经营也导致了草原生态退化（拉灿，2005）。而且，由于草原生态系统失衡严重，大面积的沙化退化使牧民生活条件恶化而导致贫困，阻碍了牧区经济的发展，使大部分牧区经济发展水平低下。显然，影响牧民生计的不仅仅是草原生态退化单一因素。

从20世纪90年代开始，国家希望通过草场承包的方式避免草原“公地悲剧”的结果。但是，草原承包并没有带来预期的效果，因为承包打破

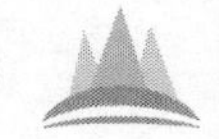

了草原的整体性，破坏了草原牧区的地方规范，加剧了草原利用的冲突。从2000年以后，国家积极地介入草原生态保护中，试图通过补贴和干预牧民的微观生产行为来保护草原生态环境。草原环境保护政策制定和实施的权力集中于中央政府，然而违规行为普遍存在，大部分地区的生态环境并没有得到改善。国家干预的失败在于国家决策的简单化和决策过程的再集中。

8.1.3 牧民的生计成本与补偿政策力度

由于禁牧、草畜平衡制定的力度加大，牧民的收入受到影响，同时还面临着修建基础设施及生活方面成本的不断提高。在对黑河流域草原管理政策的农牧民调查中得出，对未来政府草原管理的着重点方面，大部分农牧民提出首先应该“修订和完善草原保护政策”，在政策完善的基础上加大政策的监管力度和宣传教育。黑河流域草原管理存在的问题主要在于政策本身存在的问题，并不是农牧民本身缺乏草原保护意识，而是为了追求收益的不得以行为。应该在修订和完善草原管理与保护政策的基础上，加大政策的监督管理力度，这才是解决黑河流域草原管理困境的根本所在。

在对黑河流域草原管理政策执行情况的调查中，对政府补贴综合的发放状况进行统计，家庭平均获得政府的政策补贴为14 259元，其中家庭最大补贴的数值为21万元。同时，多数政府能够做到补助按时发放，占全部总数的64.6%；并且也能够做到足额发放，达到67.7%。对草原保护政策对家庭收入的影响，40.6%的农牧民认为政策使家庭收入有些增加，但也有较高比例的农牧民（26.6%）认为草原保护政策对家庭收入没有影响，甚至有16%的农牧民认为草原保护政策造成家庭收入有一定程度的下降。

因此，政府草原管理存在的根本问题在于政策本身的问题，仅仅是单一地给予农牧民补贴、禁止农牧民放牧的方式难以解决根本问题，黑河流域草原管理政策的核心在于“增加农牧民收入”，解决农民收入来源单一问题、增加当地农民收入，才是黑河流域草原管理政策的核心与解决黑河流域生态环境保护问题的根本。

8.2 西部草原可持续发展管理能力提升对策

8.2.1 明晰草原的权属，为管理提供依据

草原管理所有权和使用权是草原家庭承包责任制的基础，必须首先明晰草原所有权和使用权。要进一步明确草原使用权的内涵、范围和行使权利的方式，要有规范的法律和制度保障，并有相应的规范操作规程。赋予农牧民明晰的草原的使用权、收益权和处置权，可将草地的使用权作为商品，允许将牧户承包的草地以股份合作的形式进行自由的有偿流转。要建立和完善草地载畜量的核定制度，对于不按核定载畜量的牧户除征收草地使用费外，还要根据其超载量加收草地超载使用费。对于超载严重不自愿整改且造成草原严重荒漠化的牧户，应给予法律惩罚，并收回草原，停止承包。采取谁破坏、谁治理的原则，监督检查矿山、水电等因工程建设而破坏的草原进行植被恢复工作。

草原承包既有利于生态保护，又有利于农牧民增收。要因地制宜，加强对不同区域或不同类型草原的管理，半农半牧区草原是农牧民发展副业的生产资料，农牧业并重是这一类型的特点，在经营方式上，适度联户承包更有利于草原利用；农区草原大多没有承包到户，且人均面积较小，农民自我保护草地的意识不强，通过竞价承包，明确草地属性比较符合实际；禁牧的草原，既要保证严格的草原政策的落实，又要合理解决牧户的生计，使牧民在领取禁牧补助资金的同时要承担保护草原的责任。

8.2.2 提升草原管理的牧民参与度，加强草原社区组织建设

（1）政府相关部门主导，以社区（村）为基础开展参与式草原管理计划。在县、乡、村管理人员、技术人员和农牧民的共同参与下，经过各层次的需求评估，综合政府部门、专家和农牧民的意见和需求，特别是通过社区成员和其他相关方的参与，共同制订符合当地实际的草原管理计划。

（2）要通过各层次多方式的培训，不断提高区、乡草原管理（技术）人员的能力水平和农牧民参与管理的意识，调动牧民日常管理草地的积极性，提高保护环境意识。参与式草地管理可以促进牧民行为习惯的改变，

有利于草地生产力恢复和减轻草地退化的威胁，从而改善牧民的生活条件。

（3）引导非牧民与牧民合作经营草地，充分利用他们的资金、技术和广泛的社会关系，加强对牧民的产前、产后服务，提高畜产品效益，从而增加牧民收入，带动牧区经济发展。

8.2.3 优化牧区管理的战略布局，实施有针对性的措施

（1）积极调整优化牧区产业结构。牧区的支柱产业和优势产业是畜牧业，要利用其优势，以科技为先导，引进先进的技术和资金，大力发展以畜牧产品为原料的加工工业，推动牧区第二、第三产业的发展，优化牧业产业结构。通过产业结构的调整，使原来牧区脆弱生态系统中过度的人口压力逐步转移到非牧业的第二、第三产业中，减轻人口对草原的压力。

（2）在生态极度恶化的地区要坚决实行生态移民。荒漠化严重区或生态脆弱区的部分草原已经丧失了生态功能和生存条件，并且由于居住分散，教育、交通、就医不方便且成本较高，牧民生活十分不便。通过生态移民，将这一地区的农牧民集体搬迁至生态环境相对较好的新地方，从而使原住地生态系统得到恢复，并转变牧民生产、生活方式，有利于从根本上解决牧区生态和贫困问题。

（3）大力发展牧区生态旅游业产业。草原以其所处的地理位置，一般都具有独特的气候、植被、生态系统特征，以及社会、人文背景，具备发展草原生态旅游的优越条件，牧区大力开发旅游资源，可实现牧业人口和牧区经济向服务业的转移，同时可增加牧民收入来源和就业途径。

8.2.4 建立有效的草原管理体系，形成长效的管理制度

（1）理顺国家省市县乡多级管理部门职能及权限范围，严格草原征用审批手续，加强对占用草原的工矿企业、采砂场、旅游景点服务等场所的监管。杜绝非法开垦、乱采滥挖等破坏草原的活动，惩处恶意破坏草原生态和变更草原用途的违法行为。

（2）建立健全草原生态监测和预警体系，严格按草地生态监测技术规范，依据草原的类型、分布，建立草原资源生态监测点。加强地面定位监

测点建设，制订和完善地面调查及遥感数据分析等相关技术标准和规程。积极整合相关科研、教学、推广单位的技术及设备等优势力量，促进新技术、新成果在草原监测领域的转化和应用。

（3）积极开展草原火险预警工作，加强与气象、消防部门的合作。各草原主管部门要组织召开草原防火工作会议，落实责任，部署任务，做好预警应急工作。主要加强春季和秋季防火期督查检查，严格实施部门自查与上级抽查制度。开展草原防火信息系统应用系统的技术维护培训和草原防火物资装备储备与使用。做好草原雪灾应急管理工作，制定草原雪灾应急处理预案。

（4）加强学术研究和成果交流，积极与国内外科研院所、相关单位进行交流与合作，做到资源共享，信息互通，提高草原管理水平。

8.2.5 促进专业化合作组织发展，形成多方参与的管理格局

（1）农牧民专业合作组织的发展，不仅发展了农村生产力，还推动了农牧区经济发展，给农牧民带来了实实在在的好处。要积极创造有利于牧区专业合作组织发展的环境，要大力培育一批龙头企业，尤其是农业龙头企业，提升其对专业合作组织的带动能力。

（2）要因地制宜地发展牧区专业合作组织，组织的形式要根据区域特色多种多样。牧区专业合作组织要从实际出发，根据本地区资源优势的生产特点，围绕区域布局，充分发挥本地区支柱产业的优势，加速发展农副产品加工业，拓展产业渠道，增加附加值，在减轻草地压力的同时，实现牧民增收。要针对牧区存在的靠天养畜、粗放经营的落后生产方式，畜产品市场价格低，以及牧民收入增长缓慢的现状，引导牧民转变观念，成立合作组织、中介组织，以及草原使用权折价入股后成立股份有限公司，大力开展畜产品规模生产，提高牧民进入市场的能力，增加收入来源渠道。

（3）加快牧区专业合作组织发展。政府应加大对合作组织的资金扶持力度，拓展融资渠道，利用财政低息贴息、农业政策性贷款等方式解决启动资金和发展资金不足等问题。同时协调农村信用社和其他形式金融组织对在牧区专业合作组织的农业信贷项目给予一定优惠政策。大力培育牧区专业组织经纪人，以经纪人大户或联户为核心，组建专业合作组织。培育

农畜产品流通市场，发展市场型的农畜产品购销联合型组织。要进行农产品注册和质量认证，实施农牧产品地理名牌战略和标准化生产。

8.2.6 加强牧业生产服务体系建设，提升草原持续发展管理水平

（1）科学制订各级各类畜牧业发展规划。在进一步明确现阶段农牧业属性的基础上，合理确立近期和远期牧区产业的发展方向和发展重点，统筹设计现代畜牧服务业的功能、结构、规模和空间布局，制订科学合理的畜牧业中长期发展规划。完善和发展各种类型的服务组织，进一步加强综合服务体系建设。要建立健全良种繁育、饲料加工、疫病防治、草原建设和科技推广等服务网络，形成以国家投入为主，集体或农牧户参与的专业化服务组织。

（2）充分利用信息化网络和服务热线，通过网络、广播、电话等多种形式推进畜牧技术现代化、信息化远程服务，及时为牧户解答疑难，提高和帮助广大牧户抵御市场和自然风险的能力。以服务现代畜牧业发展为目标，建立牧业科研、推广、培训和信息咨询的服务体系。通过现有人力和物力资源的整合，做大做强服务组织，积极探索市场化运作方式。通过采取技术承包、技术入股等办法，鼓励畜牧兽医技术人员走向生产一线，在服务农民的同时，增加自身实力和收入，逐步形成形式多样、功能齐全的现代畜牧业社会化服务体系。

（3）依托资源优势，发展生态畜牧业，并以畜产品加工业为龙头，实施一体化服务网络，走全方位发展畜牧业的路子。要探索政府、企业和个人多元化投资兴办畜牧业服务组织的模式，吸收和运用各类社会资金，加大资金投入。同时各级财政要支持银行、农村信用合作联社等金融信贷机构对畜牧业服务组织的发展实行低息贷款、贴息贷款甚至是无息贷款，尽可能降低服务组织的建设成本。

第9章 西部草原可持续发展管理政策设计建议

9.1 西部草原可持续发展政策思考

9.1.1 草原管理政策制定的关键问题

近一个多世纪以来，全球不少国家都经历过草原生态退化及沙尘灾害。美国由于西部牧区调整和转变，到20世纪30年代“黑风暴”横扫西部大平原，造成几百万公顷农田废弃，上百万人口被迫外迁。前苏联也因为大面积开垦荒地，造成干旱风蚀，有1700万hm^2土地受灾。国内无论是曾经“风吹草低见牛羊”的美丽内蒙古大草原，还是青藏高原牧场，都出现了不同程度的退化现象。

草地退化是草地生态系统的退化，其后果表现在各个方面，其中最直接、最易为人们看到的是草地植被的变化。严重退化的草地，其植物群落的高度、盖度明显下降。植被变化的另一个表现是植物群落组成的变化，在家畜的过度啃食条件下，不耐牧的植物显著减少，而耐牧的植物则被保存下来，其结果导致退化草地由低适口性的植物所组成，这也是退化草地的最终类型都可能是由耐旱耐牧的植物所组成的原因。

草原退化总的表现是草地衰退，产草量降低，有毒有害及劣等草滋生，风蚀沙化，水土流失，土地盐碱化，鼠虫害猖獗。草地的退化、沙化、盐碱化这“三化”加速了水土流失，恶化了草原气候，形成了恶性循环。综

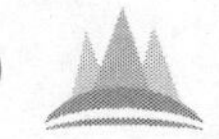

合文献资料的汇总研究，分析造成草原资源锐减、草原生态不断恶化的原因，不同研究背景的人会提出不同的观点。归纳起来主要在以下几个方面：气候论认为，气候变化是草地全面退化的基本原因，主要是干旱对草原生态退化的影响是主要的。政策论认为，农业政策与环境政策的不协调是导致草原退化的主要原因。牧区强调的“以粮为纲”等口号和政策是造成草原生态退化的政策原因；其次，草地承包落后于牲畜划分到户，草地没有实行有偿使用，草原建设管理缺乏、草原监管低效也是造成草地退化的原因。还有人认为政府采取的围栏、舍饲、畜种改良、移民等措施犯了方向性的错误。有关政策的制定没有考虑不同地区牧户生态与经济条件的多样性，对草原退化起了重要作用。

干旱是全球化、大面积草原退化的主要因素。全球受温室效应的影响，温度升高，雨水减少，干旱频繁发生，尤其是春季高温、干旱、大风等，对草原退化有重大影响。水分缺乏不仅影响到牧草的正常生长发育和产量形成，而且容易引发草原病虫灾害的发生。

人类的活动和强烈的政策干扰也是草原退化的重要因素。草原的退化很大程度上是因为传统知识被削弱，如定居导致草地利用不均衡，承包和围封导致草地破碎化，农业和矿产业的开发导致草原沙漠化。长期以来，我国对于牧业生产的考核强调的是牲畜头数的总增长率或存栏头数的净增长率，对于畜产品质量没有硬的指标，由此导致各地片面地追求牲畜头数的增长，草原载畜量上升，打破了畜草之间内在的平衡。超载过牧导致草原退化的演变是个渐变过程，单位面积牲畜数量增多，单位面积的可食性牧草被牧食的就多，使得维持牧草再生的草籽数量不足，导致可食牧草产量下降，毒草和杂草增多。这种局面若得不到控制，任其发展将出现恶性循环的局面，即单位面积上可食牧草减少而牲畜量增加，牲畜增大觅食范围和频次，一方面加重了对草地土壤结构的物理性破坏，另一方面土壤的破坏反过来会限制牧草的生长，这种恶性循环便使草原逐渐贫瘠、退化沙化。在我国北方草原沙化的成因中，超载过牧是居第二位的重要因素。

另外，草原地区开垦不仅会使被垦的耕地沙化，还能诱发周围草原的沙化。耕作破坏了草原植被，松散了生草土层，裸露松散的沙质土地在干旱的风沙中极易受风蚀，疏松的细沙土随风而起，成为沙丘的物质来源。

过多地开垦不仅缩小了草地面积、增加了草地的牲畜负荷量，而且又引起草地植被的退化。这种连锁反应使开荒区草地变成了沙地。

对于草原退化的机理研究中，任继周（任继周等，2011）指出：草原退化归结为动物生产系统和植物生产系统之间的耦合，以及在系统耦合过程中所发生的系统相悖。也就是系统运行的时间相悖、空间相悖和种间相悖。一是时间相悖，在自然状态下，动物生产的能量动态年度波动较小，而植物生产的能量动态年度波动较动物波动大。植物生产旺季，植物有机物的盈余部分，不仅浪费了草地资源，还因为对于草地的利用不足，放牧过轻而使草地植被成分与土壤变坏，草地资源受到损害。二是空间相悖，一方面是地区之间的空间相悖，就是不同地区之间，一定草地面积上动物负荷量的失调。另一方面是季节放牧地的空间相悖，冷季放牧地以较小的面积承担较长时间的放牧，负荷畸重；而暖季放牧地以较大的面积，承担较小的放牧时间，负荷畸轻。三是种间相悖，关键是生态场和生态位的合理格局。对于草地植物学组成来说，结构良好的天然群落，或不同品种合理混播的草地，可以创造良好的生态场，以提高牧草产量，改善牧草品质，抑制病虫害，保持草地肥力，延长草地寿命。这三个方面，时间相悖居于主导地位，这是植物生产系统和动物生产系统两者的节律相差悬殊所致，是最根本的相悖。

总之，由于全球气候变化、降水量的减少及牧区人口的不断增加，草原退化现象日益严重。在干旱半干旱地区，不同区域的草地管理和利用的政策调整要依据当地的自然气候特征，不能简单地复制其他地区的经验和方法。应当充分吸纳当地牧民应对气候和降水变化等方面的传统知识，制定出因地制宜的管理策略。

9.1.2 草原管理政策制定的根据

20 世纪 90 年代由于草原的大面积退化，执行严格的承载力管理成为草原治理的重要手段。然而，政府花费了大量的财力、物力和人力用来监督和管理超载过牧，但没有达到理想的预期效果。究其原因是政策的制定者理想地认为，只要将牧民饲养的牲畜控制在可计算的承载力范围内，就可以很方便地管理草地以达到永续利用草地资源的目标，这是将草原上的

“人-畜-草”关系简单化和标准化所产生的理论。

事实上，草原不是植被在数量意义上的覆盖率，载畜量也不是一个恒定值，而是一个在微观层面上的动态变量。干旱半干旱生态系统具有可变度高、不可预知的特点，传统管理理论支持的“承载力”和“载畜量”两个主要的草原管理工具在干旱半干旱地区的草原管理中存在诸多不适应性。国内外许多学者对草地载畜量进行了研究，以往大多数研究以50%的牧草可利用率计算载畜量。然而在广大的干旱半干旱区草地上，由于自然生态条件的脆弱性和不确定性，当牲畜放牧破坏了地表植被覆盖以后，很容易造成草地退化和荒漠化。上述基于草地牧草生产力50%的草地载畜量，已被不少研究结果所否定。美国土地管理局研究发现，许多地方的放牧强度在载畜量的允许范围内，可是草地却在不断退化，一些草地即使在轻牧状态下，也会毁坏植被群落的健康。

目前，大多数的政府官员或牧区管理人员对于干旱半干旱地区草原管理的重点在于保护和恢复多年生木本植物，使其生长到适当或最优状态，以获得最大的地上生长量。而一些短生植物、一年生植物、两年生植物的物种价值，在很大程度上被管理部门所忽视了。许多一年生或两年生的植物，在一年中的大部分时间里，作为藏于地下的潜在生物量以种子库的形式存活，其可持续性和可利用价值对草地具有极为重要的意义。事实上，一些植物学家和生态学家的研究表明，当有良好降水时，虽然放牧的牲畜数量超过了牧区理论上的载畜量，但由于大量繁殖的一年生或两年生植被，放牧在草地退化过程中仅仅扮演了一个微不足道的角色。

干旱半干旱地区生态系统有很强的自我恢复能力，适度的放牧不仅能够促进植物尤其是草的分蘖，而且能增加植物的蛋白质合成和提高植物的适口性和数量。由于牧草刺激的再生机制，放牧两周时，草地生产力增长一倍多，但随着放牧强度的加大，草地生产力则迅速下降。

实验表明，在草原退化严重地区实施全年禁牧，有助于植被的恢复，其改善的主要指标是草的产量、高度、盖度，有时加上草原的多样性。短时期采取禁牧措施，草原可以得到某种程度的恢复，但是从长时间来看，没有牲畜的采食，草原将会退化。禁牧后草原的恢复呈抛物线，经过短时间恢复后开始退化，适度的采食是草原恢复的重要条件，禁牧乃至严格的禁牧对草原的影响是负面的；其次，现有的实验都只计算了剪割的草产量，但是在非禁牧区，除了剪割的草产量以外，还有牲畜采食部分没有被计算在内，所以现有的研究是不完整的。

另外，中国草地资源地域广阔，并明显存在季节性不平衡。具体表现在季节牧场面积的不平衡、牧草营养的季节不平衡、牧草生长的季节不平衡、草地载畜能力的季节不平衡等方面。由于草地资源的季节不平衡，在靠天养畜的情况下，就出现了家畜“夏壮、秋肥、冬瘦、春死”的严重影响牧区发展的问题。这一情况在广大牧区具有普遍性，已成为畜牧业发展的突出问题。

不可否认，草地的承载力是有一定的界限的，超过临界值，草地就会由于过度消耗而产生退化。承载力管理是草地管理政策的进步，有利于草地的可持续发展。但承载力管理有一个空间尺度的制约，一方面，承载力计算结论受草地、降水、自然灾害等多种因素影响，同时也有研究表明，以牧户为单位的承载力管理存在多种问题，不能有效地管理和实施。另一方面，在大尺度的空间里，承载力管理是对较大范围的相对稳定的系统内牲畜数量实行总量控制。由于畜群可以通过移动来适应降水变化引发的牧草产量分布的变化，从而可以有效地避免植被稀少的地方承受相对过多的牲畜量，也可避免植物量丰富的地区没有被发现或没有被充分利用，从而实现真正意义上的草畜平衡。

9.1.3　草原管理政策制度的设立

草畜双承包责任制是指在对牲畜推行牲畜作价，户有户养，对草地实行“草地公有，承包经营”的办法。一方面是牲畜承包，主要推行牲畜“作价承包，现金提留”和“作价卖给牧民，分期偿还”的办法；另一方面是草地承包，把草牧场使用、管理和建设的责任制承包给牧户，并且保证承包期不少于20年。草畜双承包责任制设计的初衷是调动牧民生产积极性，实现草地保护和建设养畜两个目标。20世纪80年代初，农区大范围推广家庭联产承包责任制已证明承包制可以大大提高农民的生产积极性，促进农业生产的发展。在这样的背景下，草畜双承包责任制就被认为可能通过承包牲畜和草地明确牧民的责、权、利，牧民享有畜牧业经营收益的剩余

索取权，这样就可以激发牧民的生产积极性。

草畜双承包责任制实现草地保护目标的主要手段就是承载力控制，通过控制牲畜数量，使得一定区域和时间内通过草原和其他途径提供的饲草饲料量与饲养牲畜所需的饲草饲料量达到动态平衡。草畜双承包责任制中草地管理理论模型就是草地演替模型，放牧密度按照一定的比例关系影响着植被演替过程，而植被结构也相应地发生连续的和有方向的变化。放牧率越低，草地条件就越好；放牧率越高，草地条件就越差。草畜双承包制并不是一个孤立的政策，它是由一系列相互支撑的政策措施组成的一个综合的草地管理和畜牧业发展的政策体系。它不仅包括牲畜和草地双承包这种产权制度的变化；而且还包括随之引发的一系列畜牧业生产方式和草地管理方式的变革，包括定居定牧、舍饲圈养、人工饲料地的种植和草地承载力的管理等。

实施草畜双承包在一定时期对畜牧业的发展产生了积极意义。一是在牲畜饲养方式上，摆脱传统的“靠天养畜”的生产方式，实现由自由放牧向舍饲圈养的集约化经营，增强了抵御自然风险的能力。二是棚圈建设和冬季打贮草在一种程度上提高了畜牧业生产能力，减少了牲畜的冬春死亡率，提高了牲畜的成活率，减少了大起大落的“夏壮、秋肥、冬瘦、春亡”的生产性波动。三是从草地承包到户基本上制止了抢牧乱牧、无序竞争的局面，在草原上乱砍滥伐、乱开荒、乱挖乱采现象得到遏制，牧民不再盲目发展家畜头数，而将注意力放在提高家畜的出栏率和商品率方面，增加了牧业收益。

草畜双承包责任制实施后，对草原生态、牧民生计和草原文化方面也产生了诸多不利的因素。归纳起来表现在三个方面：一是在牧区游牧放牧制度彻底地被定居定牧所取代，草地被越来越多的围栏分割成细小斑块，不仅阻止了牲畜的自然流动，而且也限制了野生动物的迁徙。牲畜被限定在固定面积的草地内放牧，在相同的地方连续啃食草地，草地没有休息和恢复的时间，是造成草地退化的重要原因之一。同时，草地围封使得植被结构变得单一，减少了生态系统内的生物多样性，不利于天然草原的生物多样性演化，不利于基因资源的相对稳定性，不利于天然草原处于稳定的演替状态。二是草畜平衡的管理模式不但没有降低对草地的利用，反而会刺激牧民更多地将自然草地生态系统转变为人工草地生态系统，提高产草量，从而加大对草地的利用强度，威胁草地生态系统的生态服务功能。从畜牧业经营角度来看，多年生牧草对牲畜采食和践踏具有较强的耐受力，能够很快再生和恢复，因此其载畜能力也远远高于一年生植物。三是管理难度大，管理成本高。针对散居在广大牧区的数以百万计牧民，实现草畜的动态管理和检查监督，其成本难以预计，且监管困难。

9.2 西部草原可持续发展管理政策设计

9.2.1 明确政策对象及提高政策针对性

1. 政策服务对象

草原生态补助奖励政策实质上是对各相关方利益的调整，利益相关方的合理界定对生态奖补政策的完善和实施都具有重要的意义。在对于生态环境的保护和利用领域，利益相关方都可以分为受益者、受损者和破坏者三类。为了实现三者的均衡，生态奖补政策的实施原则是受益者付费，受损者获得补偿，破坏者应该缴纳污染费。受益者可以分为当代受益者和后代受益者，当代和后代受益者可以由政府来代表，因为政府代表了全体人民的利益，例如，草原禁牧的受益者可以是牧民的后代。受损者可以分为直接受损者和间接受损者，例如，为了恢复草原生态环境而被迫移民的牧民，就是直接受损者，而为了实施草畜平衡，因为要减少牲畜头数而产生经济损失的牧民可以转化为间接受损者。而对于破坏者，目前在政策的制定上，超载过牧被认为是主要的破坏行为，所以限制了牧民的经济发展，强调生态环境的恢复和保护。但是，工矿企业对矿产资源的开发以及生态旅游的发展，破坏了一部分草原的生态环境服务功能，尤其是工矿企业对当地其他没有被占用草原的牧户造成了环境污染，这些主体并没有为其破坏行为缴纳相应的费用，因此，应该建立生态补偿税收机制，对这些破坏草原生态环境的行为主体开征生态补偿税，并且这一部分税收要做到专款专用，用于生态保护建设资金。

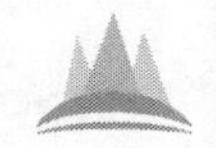

2. 政策针对性

实施草原生态奖补政策是保护草原生态环境，促进畜牧业可持续发展的重大举措。不过，要明确行之有效的政策必须具备一定的稳定性，保证其能在一定的时间内连续有效地发挥作用。虽然在实地调研中，草原生态奖补政策在如何协调草原生态环境保护、牧业发展、牧民生活的关系方面还存在问题，但是不能因此对草原生态奖补政策进行否定，对其落实与稳定产生动摇。因此，我们要提高草原生态奖补政策的持续性和有效性，并且努力在保护草原生态环境与牧民生计中找到平衡点，通过行政手段将其转变为基于市场情况的手段，运用市场机制来最大限度地解决环境问题。

9.2.2 推进草原生态政策的立法

目前，我国已经出台了《中华人民共和国环境保护法》《中华人民共和国水土保持法》《中华人民共和国草原法》等与草原生态环境保护相关的法律，但是相较于发达国家保护生态的法律法规，我国还没有制定规范的生态补偿制度。我国的生态补偿立法要远远落后于管理实践，所以在处理生态环境问题时缺乏有效的制度支持，补偿也存在一定的随意性和应急性，对利益主体也没有明确的界定和规定。可以在条件成熟时制定《生态补偿法》，通过立法，将生态补偿纳入公共财政，采取多种形式的补偿方式，如政府直接补贴、建立生态环境保护项目及地区间横向经济补偿等。

9.2.3 建立配套的资金支持政策

首先应加大中央财政投资力度，即在财政可以负担的情况下，不仅要考虑到草畜平衡，还要考虑牧民生活成本，提高草原生态奖励补助标准。此外，在转移牧民的技能培训、子女就学及社会保障等方面继续加大投入力度。另外，应加大对牧区金融经济结构调整，尽快对牧区的农村信用社实行大力度改革，在中央和自治区政府的支持下，明确信用社的产权，建立完善的内部治理结构，鼓励商业银行涉猎牧区经济，引导牧区发展民间金融。然后，要延长贷款期限，银行要结合牧民生产经营周期和牧民的实际用款需求，缩短项目审批时间。牧民居住分散、贷款需求多样，银行应

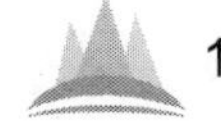

按照经济发展的不同程度考虑贷款用途，灵活制定不同的贷款政策，条件允许时可以适当地延长贷款期限。最后，要允许将草地承包权进行抵押，没有抵押担保是牧民无法获得借款的一个原因，而草场是牧民手中最珍贵的生产资料，如果可以抵押，通过信贷获得资金，对牧区牧民的生产生活水平的提高将发挥很大的作用。

9.2.4 健全牧区社会保障制度

要实现“牧民老有所养，病有所医”，真正地落实草原牧区社会保障制度，需要做到以下几个方面。首先，加快牧区社会保障制度立法，为草原牧区社会保障体系的建立提供法律支持，要明确牧区社会保障的性质、政府和企业的责任、牧民的权利和义务，保障草原牧区社会保障制度的健康顺利发展。其次，要进一步完善新型合作医疗制度，以牧民自愿参加为原则，科学地制定起付线和报销比例的具体标准，确保牧民受益，能够看得起病。此外，要健全组织机构，加强对草原牧区社会保障工作的监督管理，改变政出多门的现象，提高草原牧区社会保障工作效率。最后，要加强对牧区社会保障制度的宣传力度，增加牧民对社会保障制度的认同感。

9.2.5 实施综合性扶贫政策

牧区贫困不仅仅是经济问题，也是生态问题。国家财政对牧区扶贫的投入应随着经济的不断发展和财政收入的增加而不断增加。但是，牧区的贫困问题的解决根本上要依靠转变经济思路来解决。在内蒙古草原牧区，由于第二、第三产业滞后，牧民的知识文化水平有限，畜牧业仍然是主要的产业，政府应尽可能地创造条件，让牧民在现代社会的发展进程上获得在城市、工矿企业就业的机会，尤其是禁牧区的牧民，应该大量安排技能培训，使他们能够顺利转向第二、第三产业就业。此外，仅仅依靠政府财政作为扶贫的唯一资金来源是不现实的，应大力推动民间资本投入到扶贫项目中，鼓励引导非政府组织参与扶贫。最后要明确，草原生态奖补政策与扶贫虽然在现实中有一定的联系，但它们却有本质的区别。草原生态奖补政策的目的是保护草原生态环境和恢复已被破坏的草原环境，而扶贫的

目的是解决贫富差距，实现社会公平，实施草原生态奖补政策只是牧区扶贫的途径之一。

另外，要加大对草原牧区基础建设的扶持力度，结合当地和牧民生产生活的实际情况，在水井、围栏、暖棚建设上给予资金支持和技术帮助，真正地帮助牧民解决实际困难，减轻生产生活负担，帮助牧民实现畜牧业生产方式转型。

9.2.6 建立牧民参与政策制定的合作机制

针对草原牧区制定的政策，在制定之前应广泛地听取牧民的意见。牧民作为草原牧区传统的生活居民，他们是最有发言权以及最有实际畜牧业经营经验的群体。对草原生态奖补政策的实地调研过程中，通过对牧民的走访发现大部分的牧民表示目前的草原生态奖补政策的补贴标准偏低，认为标准定到3～4元是合理的，因为目前的补贴标准低，他们表示如果惩罚的力度不大，宁愿接受罚款也不愿意减畜。

在草原生态、经济和社会的协调发展过程中，如果不听取牧民的意见，就单方向、自上而下地制定政策是不可取的。必须要求政府聆听牧民呼声，政府和牧民合作，从而使得牧民在参与决策的过程中对于草原生态环境保护这一公共目标有共同的理解，才能够制定科学合理的符合当地牧民实际情况的补偿标准，达到草原生态环境保护和牧区经济可持续发展的双赢局面。

参考文献

奥斯特鲁姆 . 2000. 公共事物的治理之道：集体行动制度的演进 [M]. 余逊达，等译 . 上海：三联书店 .

包玉山，周瑞 . 2001. 内蒙古草原牧区人地矛盾的加剧及缓解对策[J]. 内蒙古大学学报（人文社会科学版），33(2)：93-98.

庇古 . 2006. 福利经济学（上）[M]. 朱映，等译 . 北京：商务印书馆 .

蔡虹，张相锋，董世魁，等 . 2013. 高寒牧区牧民对草地畜牧业生产现状判别及意愿的参与式调查 [J]. 草原与草坪，33（2）：66-72.

曹叶军 . 2013. 锡林郭勒盟草原生态补偿问题研究 [D]. 呼和浩特：内蒙古大学 .

陈洁，罗丹 . 2009. 中国草原生态治理调查 [M]. 上海：上海远东出版社 .

达林太，娜仁高娃 . 2010. 对内蒙古草原畜牧业过牧理论和制度的反思 [J]. 北方经济，6：32-35.

邓波 . 2004. 草原区域草业生态系统承载力与可持续发展的研究 [D]. 兰州：甘肃农业大学 .

杜发春，曹谦 . 2012. 生态畜牧业：西部草原畜牧业经济发展的走向 [J]. 原生态民族文化学刊，4（1）：118-127.

段庆伟，陈宝瑞，张宏斌，等 . 2012. 草原畜牧业发展的理论与政策探讨 [J]. 草原与草坪，32（6）：67-73.

樊恒文，贾晓红，张景光，等 . 2002. 干旱区土地退化与荒漠化对土壤碳循环的影响 [J]. 中国沙漠，6：525-533.

冯学智，李晓棠 . 2013. 我国草原法律体系的完善 [J]. 草业科学，30（1）：148-151.

冯宇，王文杰，刘军会，等 . 2013. 呼伦贝尔草原生态功能区防风固沙功能重要性主要影响因子时空变化特征 [J]. 环境工程技术学报，3（3）：220-230.

高尚玉，张春来，邹学勇，等 . 2008. 京津风沙源治理工程效益 [M]. 北京：科学出版社 .

高宇 .2013. 气候变化对黑河流域典型作物需水影响及综合应对措施 [D]. 上海：东华大学 .

巩芳，常青 . 2012. 我国政府主导型草原生态补偿机制的构建与应用研究 [M]. 北京：经济科学出版社 .

郭健，郎侠 . 2005. 西部草地生态环境现状及其功能恢复系统 [J]. 家畜生态学报，26（5)：1-3.

哈鸿儒 . 2006. 西乌珠穆沁草原畜牧业可持续发展对策分析 [D]. 北京：中国农业科学院 .

洪冬星 . 2012. 草原生态建设补偿机制—基于中国西部地区的研究 [M]. 北京：经济管理出版社 .

侯向阳，万里强，高洪文 . 2004. 我国西部草业科技发展重点的排序研究 [J]. 中国农业科学，37（4）：558-565.

侯向阳，尹燕亭，丁勇 . 2011. 中国草原适应性管理研究现状与展望 [J]. 草业学报，20（2)：

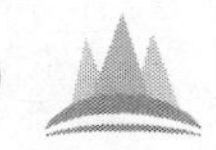

62-69.

蒋高明．2008. 以自然之力恢复自然 [M]. 北京：中国水利水电出版社．

拉灿．2005. 关于我国西部民族地区“三牧问题的思考［J］. 西南民族大学学报（人文社科版），26(5)：15-19.

李静，戴宁宁，刘生琰．2011. 西部草原牧区游牧民定居问题研究综述 [J]. 内蒙古民族大学学报（社会科学版），37（3）:7-10.

李艳波，李文军．2012 草畜平衡制度为何难以实现“草畜平衡”[J]. 中国农业大学学报（社会科学版）29（1）：124-131.

刘兴元，梁天刚，龙瑞军，等．2009. 北方牧区草地资源分类经营机制与可持续发展 [J]. 生态学报，29（11）：5851-5859.

刘兴元，龙瑞军，尚占环．2011. 草地生态系统服务功能及其价值评估方法研究 [J]. 草业学报，20（1）：167-174.

柳小妮，孙九林，张德罡，等．2008. 东祁连山不同退化阶段高寒草甸群落结构与植物多样性特征研究 [J]. 草业学报，17（4）：1-11.

曼昆．2003. 经济学原理 [M]. 梁小民译．北京：北京大学出版社．

孟慧君，滕有正，刘钟龄．2006. 论草原生态建设投资持续有效性面临的问题与困惑 [J]. 内蒙古财经学院学报，1：39-42.

蒲莉妮，郭宏远．2012. 甘南州草原畜牧业生产经营方式转变问题思考 [J]. 黑龙江畜牧兽医 ,10：18-21.

蒲小鹏，师尚礼，杨明．2011. 中国古代主要草原保护法规及其思想对现代草原保护工作的启示 [J]. 草原与草坪，31（5）：85-90，96.

秦丹．2004. 西藏草原生态经济可持续发展研究 [D]. 成都：四川师范大学．

任继周，梁天刚，林慧龙，等．2011. 草地对全球气候变化的响应及其碳汇潜势研究 [J]. 草业学报，20：1-22.

松原正毅．2002. 游牧世界 [M]. 赛音朝格图译．北京：民族出版社．

宋富民，宋鑫焱．2006. 中国西部退化天然草原机械化恢复改良的分析与探讨 [J]. 农村牧区机械化，4：57-60.

宋洪远．2006. 中国草原改良与牧区发展问题研究报告 [M]. 北京：中国财政经济出版社．

宋丽弘，唐孝辉．2012. 内蒙古草原碳汇经济发展的基础与路径 [J]. 中国草地学报，34(2)：1-7.

苏大学．2001. 西部草原治理与保护区划 [J]. 中国农业资源与区划，22（1）：14-15.

万政钰，刘晓莉．2010. 我国草原立法评价及其建议——以 2002 年底修订后的《草原法》为视角 [J]. 求索，8：148-150.

汪诗平．2006. 天然草原持续利用理论和实践的困惑——兼论中国草业发展战略 [J]. 草地学报，14（2）：188-192.

王建兵，张德罡，田青 .2013. 甘肃中西部干旱半干旱地区牧民对草原政策认知分析 [J]. 草地学报，21（1）：11-17.

王启兰，王长庭，杜岩功，等．2008. 放牧对高寒嵩草草甸土壤微生物量碳的影响及其

与土壤环境的关系 [J]. 草业学报，17（2）：39-46.

王世金，焦世泰，王淑新 . 2010. 西部生态安全建设与草原生态旅游发展研究 [J]. 中国草地学报，32（6）:101-104,109.

王文颖，王启基，王刚 . 2006. 高寒草甸土地退化及其恢复重建对土地碳氮含量的影响 [J]. 生态环境，15:362-366。

王晓毅 . 2009. 互动中的社区管理—克什克腾旗皮房村民组民主协商草场管理的实验 [J]. 开放时代，4：36-49.

吴传钧 . 2008. 人地关系地域系统的理论研究及调控 [J]. 云南师范大学学报（哲学社会科学版），40（2）：1-3.

吴永胜，马万里，李浩，等 . 2010. 内蒙古退化荒漠草原土壤有机碳和微生物生物量碳含量的季节变化 [J]. 应用生态学报，21：312-316.

颉茂华，塔娜 . 2013. 草原可持续发展三重能力的变权综合评价——以中国四大草原为例 [J]. 资源与产业，15（4）：118-124.

徐宜国 . 2009. 制约若尔盖草原畜牧业可持续发展的环境因素 [J]. 畜牧与饲料科学，30(4)：180-181.

许鹏 . 2004. 新疆草原资源特征与生态治理 [J]. 新疆环境保护，26：34-37.

闫志坚，杨持，高天明 . 2006. 我国西部草地生态恢复与建设的对策和战略思考 [J]. 北方经济，9：43-45.

杨邦杰，马有祥，李兵 . 2010. 中国草原保护与建设—内蒙古调查报告 [J]. 中国发展，10（2）：1-5.

杨邦杰，章力建，王晓斌，等 . 2011. 草原自然保护区的建设与管理—新疆草原自然保护区调研报告 [J]. 中国发展，11（1）：1-5.

杨理 . 2010. 中国草原治理的困境：从“公地的悲剧”到“围栏的陷阱” [J]. 中国软科学，1：10-17.

杨理 . 2013. 草畜平衡管理的演变趋势：行政管制抑或市场调节 [J]. 改革，6：95-100.

杨立华，欧阳赟 . 2012. 策略人和策略决策模型：草原管理的智能体模拟分析 [J]. 中国行政管理，8：106-110.

杨立华 . 2011. 多元协作性治理：以草原为例的博弈模型构建和实证研究 [J]. 中国行政管理，4：119-124.

余永定，等 . 1999. 西方经济学 [M]. 北京：经济科学出版社 .

张倩 . 2013 构建基于社区的草原管理等级框架：以内蒙古荒漠草原的一个嘎查为例 [J]. 学海，1：64-72.

张苏琼，阎万贵 . 2006 中国西部草原生态环境问题及其控制措施 [J]. 草业学报，15（5）：11-18.

张新时，唐海萍，董孝斌 .2004. 内蒙古草原的困境与可持续发展之路 [C].2004 年中国西部论坛会议论文 .

张志强，徐忠民，王建，等 . 2001 黑河流域生态系统服务的价值 [J]. 冰川冻土，23（4）：360-366.

赵锦梅 . 2006 祁连山东段不同退化程度高寒草地土壤有机碳储量的研究 [D]. 兰州：甘肃农业大学 .

中华人民共和国农业部畜牧兽医司，全国畜牧兽医总站 . 1996. 中国草地资源 [M]. 北京：中国科学技术出版社 .

周立，董小瑜 . 2013 .“三牧”问题的制度逻辑—中国草场管理与产权制度变迁研究 [J]. 中国农业大学学报（社会科学版），30（2）：94-107.

Bai Y F, Han X G, Wu J G, et al. 2004. Ecosystem stability and compensatory effects in the Inner Mongolia grassland[J].Nature, 431:181-184.

Chadden A, Dowksza E, Turner L. 2004. Adaptive Management for Southern California Grasslands [D]. Santa Barbara: University of California.

Hao Y B, Wang Y F, Cui X Y. 2010. Drought stress reduces the carbon accumulation of the Leymus chinensis steppe in Inner Mongolia, China [J].Chinese Journal of Plant Ecology, 24: 898-906.

Lesica P, Kittelson P M. 2010. Precipitation and temperature are associated with advanced flowering phenology in a semi-arid grassland [J]. Journal of Arid Environments, 74: 1013-1017.

Liebig M A, Gross J R, Kronberg S L, et al. 2010. Grazing management contributions to net global warming potential: a long-term evaluation in the Northern Great Plains [J]. Journal of Environmental Quality, 39:799-809.

Lu Y, Zhuang Q, Zhou G, et al. 2009. Possible decline of the carbon sink in the Mongolian Plateau during the 21st century [J].Environmental Research Letters, 4:1-8.

Piao S L, Fang J Y, Ciais P, et al. 2009. The carbon balance of terrestrial ecosystems in China [J]. Nature, 458:1009-1014.

Schuman G E, Janzen H H, Herrick J E. 2002. Soil carbon dynamics and potential carbon sequestration by rangelands[J]. Environmental Pollution, 116:391-396.

Straton A. 2006. A complex systems approach to the value of ecological resources [J].Journal of Ecological Economics, 56: 402-411.

Wang S, Wilkes A, Zhang Z, et al. 2011. Management and land use change effects on soil carbon in northern China' s grasslands: a synthesis [J]. Agriculture, Ecosystems and Environment, 142:329-340.

Wong N, Morgan J. 2007. Review of Grassland Management in South-eastern Australia [M]. Melbourne: Parks Victoria.

Yang Y H, Fang J Y, Tang Y H, et al. 2008. Storage, patterns and controls of soil organic carbon in the Tibetan grasslands [J]. Global Change Biology, 14:1592-1599.

附　　录

附录 1　农牧民调查问卷

中国黑河流域草原管理政策研究
——农牧民生产生活方式及政策调查

调查单位：中国科学院科技政策与管理科学研究所
兰州大学

问卷填答说明：请在所选择选项后的内画“√”，或者在横线上填适当的文字、数据；除特别说明外，问题回答为单选。

（一）被调查人基本信息

<table>
<tr><td>民　族</td><td colspan="10">A. 汉族　　B. 裕固族　　C. 藏族
D. 蒙古族　　E. 回族　　F. 其他：_______</td></tr>
<tr><td>性　别</td><td colspan="10">A. 男　　B. 女</td></tr>
<tr><td>年　龄</td><td colspan="10">A.20 岁以下　　B.21 ～ 30 岁　　C.31 ～ 40 岁
D.41 ～ 50 岁　　E.51 ～ 60 岁　　F.61 岁以上</td></tr>
<tr><td>文化程度</td><td colspan="10">A. 文盲　　B. 小学　　C. 初中
D. 高中（中专）　　E. 大学（大专）　　F. 硕士及以上</td></tr>
<tr><td>家庭状况</td><td colspan="10">您家有 ________ 口人，您家有 ________ 个劳动力；
您家有 ________ 口人从事农业生产，________ 口人从事牧业生产；
您家外出打工有 ________ 人，每年外出 _____ 天，打工收入 _____ 元；
您家（是，否）生态移民，移民补贴 _____ 元，您的期望值是 _____ 元。</td></tr>
<tr><td rowspan="2">农具、耐用品数量</td><td>电动自行车</td><td>机动四轮</td><td>机动三轮</td><td>拖拉机</td><td>摩托车</td><td>汽车</td><td>水泵</td><td>电视</td><td>冰箱</td><td>洗衣机</td></tr>
<tr><td></td><td></td><td></td><td></td><td></td><td></td><td></td><td></td><td></td><td></td></tr>
</table>

（二）被调查人经济状况

1. 经济收入状况

<table>
<tr><td rowspan="3">畜牧状况</td><td>您家放养 ________ 头牲畜，包括：牛 ____ 头，羊 ____ 只，其他 ________
您家饲养 ________ 头牲畜，包括：牛 ____ 头，羊 ____ 只，其他 ________
您每年的畜牧业收入是 ________ 元，包括肉收入 ________ 元，
乳收入 ________ 元，毛收入 ________ 元，皮收入 ________ 元</td></tr>
<tr><td>您家退牧 ________ 亩
您家经营草场 ________ 亩，其中：
租赁 ________ 亩，出租 ________ 亩
禁牧 ________ 亩，轮牧 / 休牧 ________ 亩
放牧场 ________ 亩，割草场 ________ 亩，饲料场 ________ 亩</td></tr>
<tr><td>您家的饲养方式是：
A. 全年全放　B. 全年全饲　C. _____ 月放牧，_____ 月饲养</td></tr>
<tr><td>种植状况</td><td>您家有耕地 _____ 亩，林地 _____ 亩，退耕 _____ 亩；
主要作物是 ________，产量 _____ 斤 / 亩；您家每年的种植业收入是 _____ 元</td></tr>
</table>

2. 牧业支出状况

种类	幼畜	幼禽	草料	打草机	铡草机	防疫	圈舍建设维护	雇工	其他 ______
支出（元）									

3. 种植业支出状况

种类	化肥	种子和种苗	地膜	农药	燃油	灌溉	维护	农用机械	雇工	其他 ______
支出（元）										

4. 生活消费支出状况

种类	食品	衣着、日用品	文化教育费用	医疗卫生费	交通通信费
支出（元）					

5. 您家收入的主要来源是什么？

A. 畜牧业　　B. 种植业　　C. 养殖业　　D. 生态补偿

E. 经商　　F. 外出打工　　G. 其他 __________

6. 您家庭年均纯收入是多少？

A.<1 万元　　B.1 万～ 3 万元　　C.3 万～ 5 万元

D.5 万～ 7 万元　　E.7 万～ 9 万元　　F.9 万～ 11 万元

G.11 万～ 13 万元　　H.13 万～ 15 万元　　I.15 万元以上 ________

7. 近 5 年您的家庭收入有变化吗？

A. 有很大增加　B. 有一点增加　C. 没有变化
D. 有一点减少　E. 有很大减少

8. 您对当前的经济收入满意吗？

A. 非常满意　B. 比较满意　C. 一般
D. 比较不满意　E. 非常不满意

（三）草原现状及管理政策调查

9. 您感觉近 10 年来本地气温的总体变化趋势是什么？

A. 有很大升高　B. 有一点升高　C. 基本没变
D. 有一点降低　E. 有很大降低

10. 您感觉近 10 年来本地降水量的总体变化趋势是什么？

A. 有很大增加　B. 有一点增加　C. 基本没变
D. 有一点减少　E. 有很大减少

11. 您感觉有必要对草原进行保护吗？

A. 非常有必要　B. 有必要　C. 没必要　D. 不关心
如果选择“D. 不关心”，其原因是 ____________

12. 您认为当地草地生态状况发生这种变化的主要原因是什么？

降水	气温	水资源	土地	旱灾	暴雨	冻灾	人口
增加 减少	升高 降低	增加 减少	增加 减少	增加 减少	增加 减少	增加 减少	增加 减少

13. 您对当地草原生态保护政策的评价是？

政策	退牧还草	退耕还林	禁牧、休牧、轮牧	草畜平衡	奖励补偿	其他 _____
是否执行						
是否合理						
政策效果						
监管力度						
生态影响						
对收入影响						

续表

政策	退牧还草	退耕还林	禁牧、休牧、轮牧	草畜平衡	奖励补偿	其他____
是否违犯						
是否受罚						

14. 您所在地区的草原生态保护补助奖励情况如何？

补助种类	补助标准	是否合理	您的建议值
禁牧补助	______元 / 亩	是 否	______元 / 亩
退牧补助	______元 / 亩	是 否	______元 / 亩
退耕补助	______元 / 亩	是 否	______元 / 亩
草畜平衡奖励	______元 / 亩	是 否	______元 / 亩
人工草地补助	______元 / 亩	是 否	______元 / 亩
畜牧良种补贴	______元 / 头	是 否	______元 / 头
牧草良种补贴	______元 / 亩	是 否	______元 / 亩
生产资料综合补贴	______元 / 户	是 否	______元 / 户
其他________			

15. 您家每年得到政府补贴 ________ 元？（能；不能）按时发放；(是；否）足额发放。

16. 您认为这些草原保护政策或措施的整体治理效果如何？

A. 非常好　　B. 比较好　　C. 一般

D. 较差　　E. 非常差

17. 您或者您周围熟人有没有因为违反草原管理政策受到过处罚？

A. 经常　　B. 偶尔　　C. 一般

D. 有　　E. 从来没有

18. 您认为违反草原管理政策的主要原因是？

A. 追求收益　　B. 管理松懈　　C. 处罚力度不够

D. 跟随别人　　E. 其他 ______

19. 您觉得草原保护政策或措施对您的收入造成了什么影响？

A. 大幅增加　　B. 有些增加　　C. 没有影响

D. 有些降低　　E. 大幅降低

20. 与 10 年前相比，您认为草原政策对生态环境的影响是什么？

A. 大幅改善　　B. 有些改善　　C. 没有变化
D. 有些降低　　E. 大幅降低

21. 您认为草原保护政策实施以后，生态环境变化的主要原因是什么？

A_1. 载畜量增加　　A_2. 载畜量减少
B_1. 超载过牧　　B_2. 退牧还草
C_1. 草原 / 林地开垦　　C_2. 退耕还林　　D. 其他 ________

22. 您希望怎样改变您的生产或生活方式来提高收入？

A. 增加放牧　　B. 外出打工　　C. 发展副业
D. 移民到外地　　E. 不改变　　F. 其他 ________

23. 您认为今后政府在草原管理方面应该着重于哪个方面？（单选）

A. 加大政策的监管力度　　B. 修订和完善草原保护政策
C. 加强宣传教育　　D. 其他 ________

24. 除调查问卷涉及的问题以外，您比较关心的问题有 ________

____
____
____
____
____

附录 2　管理者调查问卷

中国西部草原管理政策研究
——管理者政策执行情况调查

调查单位：中国科学院科技政策与管理科学研究所

兰州大学

问卷填答说明：请在所选择选项后的内画“√”，或者在横线上填适当的文字、数据；除特别说明外，问题回答为单选。

（一）被调查人基本信息

民　族	A. 汉族　B. 回族　C. 蒙古族　D. 裕固族　E. 其他：________
性　别	A. 男　B. 女
年　龄	A.20 岁以下　B.21 ～ 30 岁　C.31 ～ 40 岁 D.41 ～ 50 岁　E.51 ～ 60 岁　F.61 岁以上
行政级别	A. 正厅　B. 副厅　C. 正处　D. 副处　E. 正科　F. 副科及以下
部门类别	所属 ____________ 部门，负责 ____________ 工作
文化程度	A. 高中（中专）　B. 大学（大专）　C. 硕士　D. 博士
政治面貌	A. 中共党员　B. 民主党派　C. 群众
工作年限	A. 不满 1 年　B.1 ～ 5 年　C.5 ～ 10 年　D.10 年以上 ______
家庭状况	您家有 ________ 口人；________ 个劳动力
	您家有 ________ 口人从事牧业劳动，经营草场 ________ 亩，养殖 ________ 牲畜 您家有 ________ 口人从事农业劳动，家有耕地 ________ 亩，养殖 ________ 牲畜

（二）生态环境与社会政策调查

1. 您感觉近 10 年来草地状况跟以前相比有变化吗？

A. 有很大改善　B. 有一点改善　C. 基本没变

D. 有一点恶化　E. 有很大恶化

2. 您感觉近 10 年来草地退化主要表现在哪几个方面？

A. 面积变小　B. 种类减少　C. 劣质牧草增加

D. 土壤恶化　E. 其他 __________

3. 您认为当地草地状况发生这种变化的主要原因是什么？

降水	气温	水资源	土地	旱灾	暴雨	冻灾	人口
增加 减少	升高 降低	增加 减少	增加 减少	增加 减少	增加 减少	增加 减少	增加 减少

4. 您对当地草原生态保护政策的评价是？

政策	退牧还草	退耕还林	禁牧、休牧、轮牧	草畜平衡	奖励补偿	其他 ______
是否执行						
是否合理						
治理效果						
监管力度						
生态影响						
重要性排名						
成功之处						
不足之处						

5. 您对当地草原生态保护补助奖励的评价是？

补助种类	补助标准	是否合理	您的建议值
禁牧补助	______ 元 / 亩	是　否	______ 元 / 亩
退牧补助	______ 元 / 亩	是　否	______ 元 / 亩
退耕补助	______ 元 / 亩	是　否	______ 元 / 亩
草畜平衡奖励	______ 元 / 亩	是　否	______ 元 / 亩
人工草地补助	______ 元 / 亩	是　否	______ 元 / 亩
畜牧良种补贴	______ 元 / 头	是　否	______ 元 / 头
牧草良种补贴	______ 元 / 亩	是　否	______ 元 / 亩
生产资料综合补贴	______ 元 / 户	是　否	______ 元 / 户
其他 ______			

（三）草原管理政策执行情况调查

6. 您感觉本地政府在草原保护方面的态度如何？

A. 非常重视　　B. 比较重视　　C. 一般重视　　D. 不重视

7. 您感觉本地对草原保护政策或措施的监管力度总体情况如何？

A. 很强　B. 较强　C. 一般　D. 较弱　E. 很弱

8. 您感觉农牧民违反草原保护条例的情况如何？

A. 很多　B. 有一些　C. 比较少　D. 几乎没有

9. 您认为农牧民违反草原管理政策的主要原因是？

A. 提高收益　B. 管理松懈　C. 处罚力度不够

D. 跟随别人　E. 其他 ________

10. 您认为这些草原保护政策或措施的整体治理效果如何？

A. 非常好　B. 比较好　C. 一般

D. 较差　E. 非常差

11. 您觉得本地草原保护政策或措施对牧民收入的影响？

A. 大幅增加　B. 有些增加　C. 没有影响

D. 有些减少　E. 大幅减少

12. 您认为今后政府在草原管理方面应该着重于哪个方面？

A. 加大现有政策的实施力度　B. 修订和完善草原保护政策

C. 提高牧民的收入水平　D. 提高牧民草原保护意识

E. 加大草原保护政策的宣传力度　F. 其他 __________

13. 除调查问卷涉及的问题以外，您比较关心的问题有 ____________

____

____

____

____

Research into the Sustainable Development Management of Western Grasslands in China

A case study of the Heihe River Basin

Chapter 1

Research Background and Significance

1.1 Research Background

Against the background of global warming and the frequent occurrence of natural disasters, it has been a worldwide consensus to protect the ecological system. Grasslands, which are considered as a renewable resource, are essential for the survival and development of human beings. Prior to anthropogenic intervention, grasslands accounted for 40% to 50% of the Earth's surface. Due to the advancement of stockbreeding, this figure has been decreasing gradually. According to the statistics from the Food and Agriculture Organization of the United Nations (FAO) in 2000, the acreage of world's grasslands is about 35 million km^2, making up 26.91% of the land surface. The rational exploitation and the efficient protection of grassland resources have become an important means to both preserve the ecological environment and biodiversity, and satisfy the needs of human development. Therefore, it is global topic no one could ignore.

With a total area of 400 million hectares of native pastures, China is the second largest grassland country in the world only next to Australia. About 3.2 times the size of its arable land and 2.5 times the size of its forest, grasslands in China make up 41.7% of its land surface. As the largest terrestrial ecosystem in this country, grasslands comprise 63% of China's vegetation ecosystem, which also include farmland and forest. However, since the 1990s, the problem of water shortage has given rise to many problems facing the grassland community. Therefore, grassland eco-environment protection has received more and more attention. According to Gao, et al. (2008), massive sandstorms hit the northern part of China 5 times in the 1950s, 8 times in the 1960s, 13 times in the 1970s, 14 times in the 1980s and 23 times in the 1990s. Sandstorms are the most direct natural disaster causing grassland desertification (including degradation and salinization). According to the statistics from the Ministry of Agriculture of the People's Republic of China (MOA), around 90% of China's available natural grasslands are suffering from desertification, degradation and salinization to a different degree with an increase rate of 2 million hectares per year.

The ever-worsening eco-environment, especially the deterioration of the grassland eco-environment, resulted in severe sandstorms, jeopardizing the well-beings of local dwellers and even the people both at home and abroad. These problems have attracted the attention of the Chinese government. The Decision of the Central Committee of the Communist Party of China (CPC) on Some Major Issues Concerning Comprehensively Deepening the Reform, which was approved at the Third Plenary Session of the 18th Central Committee of the Communist Party of China (CPC), gives top priority to the development of an ecological civilization system. It points out that the comprehensive buildup of the ecological civilization system should include the systems of water source conservation, damage compensation, responsibility, environmental harness and ecological restoration. The No.1 Document of the central authorities in 2014 also gives importance to grassland protection, which leads to the resumption of the Green for Grain (GFG) Program, and the launch of a program on the development and utilization of southern grasslands and the protection of grassland nature reserves.

1.2 Significance of Research

With a total land area of about 675 million hm^2, the western region of China

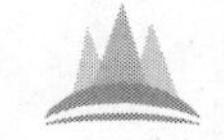

is composed of 12 provinces or autonomous regions, namely, Inner Mongolia, Xinjiang, Shanxi, Ningxia, Gansu, Qinghai, Sichuan, Chongqing, Tibet, Guizhou, Yunnan and Guangxi. The region's grassland acreage totals 331 million hm^2, accounting for ca.49.04% of its total land and 84.01% of China's total grassland area (Zhang and Yan, 2006). As to grassland harness and protection zoning, some articles (Su, 2001) divide the western grasslands into three parts, i.e. Inner Mongolia-Ningxia-Gansu temperate steppe, northwest arid desert and mountainous steppe, and alpine meadow on the Qinghai-Tibet Plateau. Regarding animal husbandry zoning, the region can be divided into four gazing areas in Inner Mongolia, Xinjiang, Qinghai and Tibet.

From a systematic point of view, the grassland ecological system is a complex ecosystem, which is composed of three subsystems, namely, the environment, economy and society. However, the important role of grasslands as a terrestrial ecosystem has not been fully recognized and highly valued. During the exploitation and utilization of grassland resources, the pursuit of immediate economic profits at the cost of natural ecological law violation gives rise to severe problems, such as the ever-worsening resource crises, land degradation and frequent occurrence of sandstorms. In the western region of China, which mostly consists of Aeolian sandy areas, soil erosion areas and areas for water resource conservation, grasslands are the main component of its eco-environment system and the foundation for the survival and development of local ethic minorities. Enhancing its eco-environmental protection and cultivation is essential for promoting national eco-environmental progress and satisfying the objective requirement of local socioeconomic development.

Western grasslands, which are extremely sensitive forefront areas for both national economy and eco-environmental conservation, play an irreplaceable strategic role in ecological safety. However, the recent decade has witnessed increasingly severe problems facing pasture regions, grassland farming and herdsmen. The grassland ecosystem keeps degrading while the cost of farming keeps growing. The living standards of herdsmen are apparently decreasing while the gap between the rich and the poor is widening. The following is suggested reasons behind the above situation. Firstly of all, the pursuit of economic interests leads to severe over-grazing in most part of western grasslands. With the extension of the practice "contracting grasslands to the herdsman households" and inappropriate grazing zoning, every inch of grasslands has been frequently utilized, which results in the grassland exhaustion and erosion. In addition, alien and poisonous species invasion causes grassland degradation. Secondly, deforestation and mining result in primitive vegetation reduction, water loss and soil erosion. Over-exploitation of crude drugs causes grassland degradation. Thirdly, locating in inland arid and semi-arid regions, western grasslands are less subject to the influence of marine climate. Local annual rainfall is 50-450 mm and groundwater is mostly 50-100 mm. Excessive deforestation leads to vanishing water conservation forests and lowering groundwater capacity, which finally accelerates the degradation of grasslands precipitation fall, snowline rise and glacier retreat (Guo and Lang, 2005). Finally, over-hunting leads to the reduction of the living areas for wildlife and vegetation, deteriorating local habitat and biodiversity. As a result, grassland ecological equilibrium is disturbed, which reduces the acreage of grasslands.

With a total length of about 821 km and covering an area of 1.429×10^5 km^2, the Heihe River Basin is the second largest inland basin in the arid areas of Northwestern China. From its water heads to the Juyan Lake, the basin goes across three various natural environmental units. Geographically locating between 96°42' and 102°00' E, and between 37°41' and 42° 42'N, it is composed of 11 counties in Qinghai, Gansu and Inner Mongolia. Qilian County in Qinghai is in the upstream, counties of Shandan, Minyue, Zhangye, Linze, Gaotai, Su'nan and Jiuquan in Gansu are in the midstream, and Jinta in Gansu and Ejina in Inner Mongolia are in the downstream.

The features of this ecosystem are similar to other inland river basins in the arid areas of Northwestern China. Over the recent 50 years, as many as 98 reservoirs had been built across the Heihe River. Land reclamation has

given rise to a large area of artificial oasis there, with tremendous agricultural economic benefits. However, due to the severe human activity, land utilization and coverage pattern have led to a profound change, causing a continuous degradation of the eco-environment. The main problems in the upstream include grassland degradation (such as grassland desertification and poisonous vegetation sprawl), reduction of natural forests and biospecies, and glacier decreasing. The midstream is mainly plagued by desertification and secondary salinization. In the downstream, major ecological problems include lake degradation and abolishment of natural river courses, which results in inland desertification expansion and oasis shrink.

Chapter 2

Literature Review on Grassland Management

2.1 Functions of the Grassland

2.1.1 Ecological function

Ecological function of grasslands, which refers to their habitats, biological properties or ecological system processes (Straton, 2006), mainly works on the grassland-surface interface. It provides living systems with the natural environment, playing a life-supporting and environment-adjusting role. As a foundation for socioeconomic development, the eco-function of grasslands includes water conservation, soil formation, erosion control, waste treatment, sand and dust retention, and biodiversity maintenance (Liu, et al., 2008). The function, as the endogenous features embodied in the social attributes of grasslands, is difficult to be commercialized, showing its indirect economic values.

Occupying 41.7% of national territory, China's western grasslands play an irreplaceable strategic role in the country's food supply and diversification, and ecological safety. However, the grasslands confront severe problems such as deteriorating eco-environment, insufficient water and soil resources, and excessive exploitation of vegetation and crude drugs. Over-grazing, irrational

exclamation and deforestation have also damaged the grasslands, their vegetation and eco-functions, which have resulted in the decreasing productivity of grasslands. Now the unit productivity of the grasslands in China's west only equals to 1/27 of that in the US.

Table 2-1 Distribution of eco-functional zones and their main ecological problems in western grasslands

	Water-conservation areas	Wind-break and sand fixation areas
Distribution in the western region	Frigid alpine deserts, pastures, meadows of the Qilian Mountains in Gansu; grasslands in the southern part of Gansu; Qinghai wetlands and upper reaches, the source region of the Yellow River and the Sanjiangyuan region; the upper reaches of the Yarlung Zangbo River, alpine steppes in the northern part of Mid-Himalayas; desert steppe of Ertix-Ulungur River, mountainous grasslands in the western part of the Junggar Basin, grasslands in the north-side and south-side of the Mt. Tianshan	Desert steppes in the central part of Gansu and Ningxia, desert steppes in the Tengger Desert, desert steppes in the middle and downstream of the Heihe River, the Mu Us Desert, frigid alpine deserts in the Northeast part of the Qaidam Basin, grasslands in Gonghe Basin, desert steppes in Altyn, bushes in the east part of Junggar Basin
Main ecological problems	Glacier retreat, snowline rising; wetland decreasing, vegetation deterioration, meadow degradation and desertification; weakened function of water, soil and biodiversity conservation	Over-grazing and over-exploitation on grasslands; severe degradation of grassland eco-system due to irrational utilization of water resources; monotonous community structure; decreasing species composition; increasing desertification, frequent pest and rodent plagues; decreasing biomass and productivity; frequent sandstorms

Data source: Wang et al. (2010)

2.1.2 Productive function

Productive function of grasslands refers to their production of consumption resources for living systems, which is the specific reflection of productive attribute of the grassland ecosystem. Acting on the grassland-creature interface, the function includes the following procedures: nutrient cycling and storage, carbon dioxide fixation, dioxygen emission, and sulfur dioxide subtraction. These functions can be commercialized, embodying its direct economic value (Wang et al., 2008).

With the total area up to 400 million hm^2, China's grasslands play an important role in carbon sink. Recent research findings in the field mainly concern the impact of meteorological factors (i.e. carbon dioxide and temperature, etc.) and management practice (i.e. grassland reclamation, grazing, mowing, fertilizing, and artificial grassland re-planting, etc.) on the carbon sink function of grasslands and their underlying mechanisms, which can be used as reference to grassland management practice in China.

1. Carbon sink of grasslands

Grasslands account for about 1/3 of the total land area on the earth. The total grassland storage of carbon is about 308 Pg, of which about 92% (282 Pg) is reserved in the soil. In low entropy biomass, the percentage of carbon is less than 10% (26 Pg) while organic carbon in grassland soil accounts for 10%-30% of the total organic carbon (Schuman et al., 2002). According to *2010 National Grassland Monitoring Report* from MOA, the total carbon storage in China's grasslands is about 42.73 Pg, accounting for 9%-19% of the total carbon storage of the world's grasslands.

Research (Piao et al., 2009) shows that carbon sequestration of China's grasslands is much greater than their carbon emissions. From the 1980s to the 1990s, for instance, they sank (0.0076+0.002) PgC each year. Carbon storage in the grasslands of the Qinghai-Tibet Plateau is about 55.9 gC/m^2 per year, accounting for 50% of the total carbon storage of China's grasslands. Thanks to the grassland replanting program (0.03 billion hm^2) and fencing enclosure campaign (0.15 billion hm^2), according to Wang et al. (2011), 0.24 PgC could be fixed each year by China's grasslands by 2020. Based upon the 44-year statistics of organic carbon in grassland soil in the US, Liebig et al. (2010) believed that the Great Plains is an important carbon sink with the total carbon sequestration rate up to 0.39-0.46 MgC per hectare per annum. Lu et al. (2009) and Wang et al. (2011) suggested that the role of grasslands as a carbon sink or a carbon source could be interchangeable, subject to meteorological factors (i.e. temperature and precipitation) and management practices.

2. Impact of climate change on carbon sink

Ren et al. (2011) found out that the main carbon sinks on China's grasslands

are those with relative high generation productivity, such as alpine meadows, temperate humid grasslands and semi-desert grasslands. Other studies (Bai et al., 2004; Lesica and Kittelson, 2010) showed that grasslands' carbon sink capacity goes up when their productivity and moisture utilization rate improve along with increased carbon dioxide concentration in the atmosphere. However, biomass is significantly correlated with precipitation in desert grasslands. Increasing precipitation would enhance the productivity of grasslands thus their capacity of carbon sink. Droughts lowered the productivity and carbon accumulation of *Leymus chinensis* steppes in the Xilin River Basin, Inner Mongolia, leading to the transfer from carbon storage to carbon sources (Hao et al., 2010). It is also reported that the capacity of carbon sink in alpine grasslands of the Qinghai-Tibet Plateau increases when precipitation goes up, while temperature is not significantly correlated with their carbon sink capacity (Yang et al., 2008).

3. Impact of grassland degradation on carbon sink

Wu et al. (2010) and Zhao (2006) revealed that organic carbon in soil was apparently decreasing when grassland's degradation became more severe. According to Wang et al. (2006), there is a critical carbon loss up to 3.80 kg/m^2, amounting to 50.87% of the organic carbon loss, when there was serious grassland degradation (0-20 cm). Studies by Wang et al. (2011) estimated that severe grassland degradation in Northern China from the 1960s to the 1990s caused a large amount (up to 1.24 Pg) of net carbon loss. Due to the harsh natural conditions and interruption of human activity, soil desertification would easily happen in the arid region, which could result in degradation procedures, including erosion, salinization and hardening of soil. It would also lead to the mineralization of organic carbon and inorganic carbon, increasing carbon emissions. On the other hand, it would cut down biomass productivity, deteriorating the soil structure and increasing the soil capacity. The percentage of inorganic soil in arid regions would amount to more than 2-5 times of organic soil (Fan et al., 2002). That is, controlling the retention of inorganic soil in arid region is of great significance to the grasslands' capacity of carbon sink.

2.1.3 Life-supporting function: regional development (addressing issues concerning herdsmen, husbandry and pasturing areas)

Besides the ecological and productive functions, the grassland ecosystem also supports the living and production activities of human beings under certain environmental conditions. It is a comprehensive supporting system combining human, pasture, husbandry and ecological environment. Acting on the interface of grass-livestock-operating-administration, this function includes livestock production, cultural heritage and leisure tourism. Some of them could be commercialized, representing their direct economic value while others may not be quantified representing their indirect economic value.

Issues concerning herdsmen, husbandry and pasturing areas have aroused great attention of the central government. On October. 12th, 2010, then Premier Wen Jiabao hosted an executive meeting of the State Council and decided to introduce a subsidy mechanism for grassland ecological protection to increase herdsmen's incomes. Starting from 2011, the Central Committee of the Communist Party of China has earmarked 13.4 billion yuan each year to help perform the task in eight provinces or autonomous regions (Inner Mongolia, Xinjiang, Tibet, Qinghai, Sichuan, Gansu, Ningxia and Yunnan) (Du and Cao, 2012).

Livestock increment and productivity reduction in western grasslands have led to a severe contradiction, which would finally cause over-grazing and grassland degradation. Most of the western grasslands grow naturally, and are subject to meteorological conditions such as precipitation and temperature. Their harvest differs greatly in different years. Especially, the quantity and temporal distribution of precipitation have direct impact on the growth of vegetation of the year and even the following year. This would be one of the factors that may go against to grassland sustainable development. When the eco-environment of grasslands has deteriorated, animal husbandry there would be vulnerable and

unstable. It is reported that average mortality of livestock in Qinghai amounted to 1.5 million each year and the productivity of its livestock was continuously decreasing (Yan et al., 2006).

Due to the natural conditions and historical reasons, socio-economic growth and science-culture development in the western grasslands have been low and backward. Especially in the recent decades, the ecological environment of grasslands has suffered from further deterioration, leading to the increment in poor people. In 1995, original value of fixed assets in Shanghai was 24.65 million yuan/km^2, 6.35 million yuan/km^2 in Beijing, 1.6 million yuan/km^2 in Jiangsu, 31 100 yuan/km^2 in Qinghai, 51 800 yuan/km^2 in Inner Mongolia, and 150 000 yuan/km^2 in Gansu. Up to 1999, about 153 counties in China's northwest were still in the state of poverty (Yan et al., 2006).

In addition, with the further promotion of herdsmen settlement projects such as ecological migration, neo-pastoral areas and nomad's dwellings, more studies have been conducted on the policy-orientation, implementation scale and methods, cultural transition and living adaptability of the projects. Li et al. (2011) pointed out that the psychological adjustment of herdsmen was the primitive challenge to the initiatives, which could also be regarded as the fatal issue to achieve perpetual settlement. Further attention should be given to issues concerning herdsmen's mental capacity of tolerance, and research into issues concerning ethnic policies, ethnic work in communities, settlement policies and its enforcement during the urbanization drive.

National laws, regulations and policies have been made to tackle issues concerning herdsmen, husbandry and pasturing areas (Table 2-2).

Table 2-2 National laws, regulations and policies to deal with issues concerning herdsmen, husbandry and pasturing areas

Year of promulgation	Laws or regulations	Issuing departments	Major content
2002	*Grassland Law of the People's Republic of China* (revised)	Standing Committee of the National People's Congress of the People's Republic of China	Breeding quantity should not exceed the carrying capacity of livestock; contracting operators should follow the rotational grazing in different zones; captive breeding in farming areas and farming-pastoral areas is advocated; changing the productive means of animal husbandry only relying on natural grazing land
2002	*Opinions of the State Council on Strengthening the Protection and Development of Grasslands*	The State Council	Protective system for primary grasslands; livestock-forage balance system; rotational grazing in different zones, grazing moratorium and grazing ban according to different situations; fencing and hydrological infrastructures in pastoral areas; captive breeding
2002	*Opinions on Strengthening the Development of Animal Husbandry*	Ministry of Agriculture	Rational utilization of grassland resources; forage–livestock balance; rotational grazing in different zones; efficient husbandry; rotational farming in farming-pastoral areas
2005	*Animal Husbandry Law of the People's Republic of China*	Standing Committee of the National People's Congress of the People's Republic of China	Building infrastructures in pasture areas, such as fencing, hydrological construction, grassland improvement, forage base, etc. to improve eco–environment
2007	*Opinions on Promoting the Sustainable and Healthy Development of Animal Husbandry*	The State Council	Full implementation of livestock–forage balancing; natural pasture preservation; forage base development; introduction of advanced breeding techniques; enhancing the capacity of husbandry
2011	*Opinions on Promoting Sound and Rapid Development of Pastoral Areas*	The State Council	Enforcing grazing-banning, grazing-off and rotational grazing step by step; reducing the carrying capacity of livestock in the natural pastures to realize the balance between livestock and forage; development of water-saving irrigating forage bases; grazing-banning rather than breeding-banning

Data source: www.gov.cn; From related website of the State Council and the Ministry of Agriculture.

2.2 Theoretical Research on Grassland Management

Currently there are four theories about grassland management: systematic development theory, localized development theory, substitute development theory and nomadic development theory (Table 2-3).

Table 2-3 Basic theories on grassland management

Related theories	Major ideas	Leading representatives
systematic development theory	Including four-production-level theory and the system-coupling theory, this theory tries to realize the optimization and development of resources using the interface theory	Ren Jizhou
Localized development theory	Stressing local experience and resisting strategies inappropriate to local development, this theory emphasizes the full observation of the development pattern of the animal husbandry and is cautious in making impact on the agricultural eco-system	Da Lintai, Wang Xiaoyi, et al.
Substitute development theory	By altering the mode of production and rehabilitation in pastoral areas, production in pastoral areas will be transferred to other areas, which will be beneficial to the grassland recovery	Zhang Xinshi, Xu Peng, Jiang Gaoming, et al.
Nomadic development theory	A full nomadic restoration will be beneficial to the recovery and sustainable development of grasslands. Cultural desertification is the origin of the desertification of grasslands and sand storms	Matsuhara, et al.

Data source: Duan et al. (2012)

Comparatively speaking, Xu and Zhang's perceptions on grassland recovery and reconstruction have some similarities in essence. Fundamentally speaking, Ren's perception on the system-coupling theory is similar to the ideas of Xu and Zhang, which gives importance to resource integration and conservation, and neo-resource exploitation. Wang pointed out that vegetation and groundwater exploitation for the sake of grassland recovery will only make severe impact on the grassland ecosystem. Therefore, damage to groundwater while cultivating artificial pasture may exert profound influences on the grassland ecosystem.

However, grassland recovery and reconstruction is a cross-disciplinary issue involving different theories of ecology, resource and environmental science, and economics.

2.2.1 Theory of public goods

According to microeconomics, the consumption of public goods has two characteristics, competitiveness and exclusiveness. Economists have different opinions on the categorization of goods. Mankiw (2003) classified the goods into private goods, public goods, common resources and natural monopoly. Vincent Ostrom and Elinor Ostrom (2000) described the goods as private goods, public goods, charged goods and common-pool resources. Yu et al. (1999) sorted the goods into pure public goods and quasi-public goods according to their degree of non-competitiveness and non-exclusiveness. Compared to private goods, Paul Anthony Samuelson regarded non-competitiveness and non-exclusiveness as the two basic characteristics of public goods.

Ecosystem services provided by the ecological construction of western grasslands are a sort of complicated public goods. Ecological construction of grasslands is to recover and preserve the degraded ecosystem services with the final goal to achieve and ensure the sustainable supply of the ecosystem services. According to previous studies on the definitions of public goods, non-competitiveness of grassland ecosystem consumption can be explained as the ecosystem services that are provided to certain consumers without sacrificing others' consumption of this ecosystem services. That is, the marginal cost of one unit of an increment on ecosystem service equals to zero. For instance, carbon sink, as one of the ecosystem services of grasslands, benefits the whole international community on the aspect of lowering the threats from global warming. Moreover, the non-exclusiveness of grassland ecosystem consumption can be explained as the ecosystem services that are provided to the consumers who have been charged will by no means exclude those who have not been charged.

Since grassland ecosystem services have the two basic characteristics, non-competitiveness and non-exclusiveness, therefore, consumption and exploitation of grasslands would also confront the same problems as other public goods do: "tragedy of the commons" and "taking a free ride". If we allow the market to allocate a certain sort of ecological resources with the characteristics of non-exclusiveness, then the "invisible hand" may cause "market failure". This would also lead to over-utilization of public resources, which would definitely deteriorate the ecological environment and resources. The consequences would be irreversible. For the supply of public goods, therefore, economists and policy-makers encourage the involvement of the "visible hand", which is the public ownership market, to compensate the groups who have sacrificed their self-interest for environmental protection by setting up some kind of compensation

mechanism. This would improve the welfare of the whole society. Besides, non-competitiveness and non-exclusiveness of public goods are not invariable. A public good would be commercialized and supervised as a private good due to the ownership clarification, technological improvement and perfection of laws and regulations.

2.2.2 Theory of externalities

Externalities, one of the widely discussed and utilized issues, are one of the important domains in classical economics. Marshall published *Principles of Economics* in 1890 that firstly launched the concept of externalities. However, Pigou, as one of the generators of welfare economics, was the first to launch the systematic theories of externalities. Pigou (1971) proposed an explanation for the essence of externalities. According to the perspective of welfare economics, externalities are regarded as the external influence from one economic entity to the other. Such influence is not achieved by the price mechanism, which leads to the variance between social costs and private costs. Regarding the effect of externalities, it can be localized into a certain area, but also can be globalized. It can also be positive (when private benefits of economic entities are less than social benefits) or negative. The restoration of the grassland ecosystem and its services has positive externalities. The improvement of local ecological environment, would maintain the integrity and interrelation of the ecosystem over larger territory, even enhance the welfare of human beings.

However, if no action has been performed to address environmental disruption, the ecosystem may be deteriorating, which would finally cause irreversible results. From the aspects of externalities, if benefits of ecological protection are free of charge and the damage to the victims has not been compensated for, it would hardly lead to the optimal allocation and utilization of ecological resources. Therefore, actions should be taken to correct or alleviate the negative externalities, that is, to internalize the externalities. Theoretically speaking, "Coase Theorem" and "Pigou Theorem" have been regarded as two potential solutions. For "Coase Theorem", it concentrates on private trades. If the transaction cost is low enough and not related to property rights, entities of individuals, communities and even the super-state would continue the transaction over their property rights till the Pareto Optimality of ecological products and services. For "Pigou Theorem", it is more related to policy-making regarding, for instance, taxation and subsidy. By means of government intervene, positive externalities are subsidized to compensate the costs of producers and deserved profits. This would increase the supplies of externalities as well as the welfare throughout the society.

2.2.3 Theory of human-land relationship

As a fundamental theorem of human geography, human-land relationship theory concerns the inter-relationship between human activities and geographical environments. Based upon a certain territory over the surface of the Earth, human-land relationship is a dynamic structure formed through inter-connection and inter-correlation between human and land (Wu et al., 1997). It requires a constant dynamic coordination among population, resources, environment and social economic developments (PRED for short) over a certain territory (Mao, 1995). More studies have been conducted based upon the theory.

In terms of grassland restoration, it is important to note that socioeconomic development requires the exploitation and utilization of the eco-environment system. However, in order to balance human activities with land resources, human beings as well as society have to adapt their behaviors to the current situation and future demands correspondingly. Compensation should also be made for ecological disruption. Based upon the above understandings and related actions, the coordinated development among PRED would be achieved.

2.2.4 Theory of sustainable development

In 1987, the United Nations launched a research report, *Our Common*

Future. It firstly defined sustainable development, that is, "development that meets the needs and aspirations of the present without compromising the ability of future generations to meet their own needs". This definition contains two aspects. One is concerning the inter-generational equity (ecological justice, or environmental justice), which means that the human activities of the present generation may also influence the future generations, especially in terms of environment and resource reservation and utilization. It refers to "a plan on reserving the total production capacity for the future generations". The other concerns the constraining impact of ecological environment on socioeconomic development. All economic activities should be ecologically friendly so as to impose a restriction on pollutant emissions and ecosystem services.

Economists and ecologists have different perspectives over the issues of sustainable developments. Ecologists pay more attention on the balance and the harmonious advancement of human and nature. Environmental background should be taken into serious consideration during human development. Natural resources are classified into values in use and nonuse. Taking air purification as an example, the function has been regarded as the values that cannot be substituted by the productive capital, human capital and social capital. Economists emphasized the maintenance and improvement of living standards. Only when the present generation ensures that the capital stock for the future generations is no less than the current ownership, then natural resources could be substituted by capitals of other forms (productive capital, human capital and social capital) for the purpose of production and consumption.

Based upon the above discussion, economists and ecologists have different attitudes toward resource and environment protection and exploitation. Ecologists concern more on ex ante precaution while economists focus more on the ex post compensation. Ecologists can hardly reach a consensus with economists on the idea of "polluter pays". Ecologists believe that ecological disruption, under most circumstances, is irreversible, and natural capital, to some extent, is irreplaceable.

2.3 Research on Grassland Management Countermeasures

From the perspective of welfare economics and institutional economics, Yang (2010) reveals the dilemma now facing China's grassland management, which is changing from "tragedy of the commons" to "trap-barrier system". Due to the imperfect institutional and administrative conditions, Yang suggests, grassland degradation has resulted from the inevitable equilibrium achieved by rational economic men who are striving to exploit grassland and water resources without renting. However, the solution to this problem is not encouraging the household fencing, but the more diversified self-governance by local communities.

Pu et al. (2011) reviewed ancient theories and ideas regarding natural grassland management from the perspective of historical course (Table 2-4). Their advices include enhancing legal enforcement, increasing investments, and improving the social economic development.

Table 2-4 Measures on grassland protection of ancient China

Nomadic tribes/minorities	Measures
Mongolian	Several taboos and punishment rules were put forward, including banning on deforestation and grassland-digging, and punishment on ignorance of fire-extinguishing
Tibetan	Due to the religious reasons and the psychological causations (fear, respect and awe), Tibetans had an ecological perspective on coordinating human beings and the nature
Dangxiang (the Western Xia regime)	A large proportion of legislations in the Western Xia regime were related to animal husbandry. All the national punishments and rewards were conducted according to farming, grazing and battle achievements

Data source: Pu et al. (2011)

Yan et al. (2006) put forward detailed technical means to promote ecological restoration and improvement of grasslands, including reform on the production and operation of husbandry, the development of water-oriented forage bases and artificial grasslands, and the vegetation restoration and reconstruction. Besides, it is also necessary to strategically strengthen grassland resource management. Conducting evaluation on resource protection policies and ensuring the leadership responsibility would also be essential for the grassland

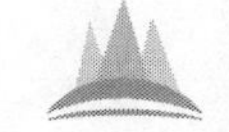

management. People's awareness on resource protection could be enhanced by population control in pastoral areas. Song and Song (2006) suggested that mechanization would be an important means to improve the degraded western natural grasslands in China, both in terms of output and quality. It would solve the current situation of over-grazing and insufficient forage grass. Chen and Luo (2009) looked into the following problems challenging grassland governance. First of all, the level of compensation paid to herdsmen and farmers is pretty low and the compensation mechanism is incomplete. Secondly, production investment is severely insufficient and credit support is urgently needed. Thirdly, the conflict between grasslands and grazing is still rigorous since the long-term mechanism behind forage supply is insufficient. Moreover, the supply of public services and infrastructures is insufficient and could not satisfy practical demands. Also, follow-up industries are lagging behind, checking the increase of farmers' and herdsmen's income. Gong and Chang. (2012) discussed grassland environmental Kuznets curve based upon environmental Kuznets curve (EKC). They also conducted theoretical analysis on the influence of government-oriented grassland ecological compensation mechanism on the grassland EKC. It includes the horizontal and vertical improvement. As to the former, government-oriented compensation could cause the curve to reach the peak when the income level is pretty low, which would accelerate the restoration of grassland eco-system and the balance development of ecology, economy and society. As to the latter, the introduction of government-oriented compensation mechanism and grassland eco-system improvement would decrease the positive inflow of entropy and increase its negative inflow, which would lower the peak altitude of grassland EKC so as to achieve the sustainable development of grassland eco-economy.

Other corresponding countermeasures have been proposed to address the three issues concerning herdsmen, husbandry and pasturing areas. Du and Cao (2012) suggested that ecological husbandry is the direction of, and solution to, the economic development of western husbandry in the future. They argued that ecological husbandry should not adopt a uniform method, and different productivity modes are needed according to productivity level and regional characteristics.

Regarding adaptive management, Chadden et al. (2004) came up with a detailed research plan to deal with such problems as alien species invasion, over-grazing, climate change and urbanization. They examined the current situations of grassland ecosystem comprehensively based upon literature review and data analysis, and offered advices to the ecologists and administrators. They also set up serial forecasting models and experimental pilots. In Australia, theories and methods on active adaptive management has been installed in the grassland ecosystem of Southeast Asia. In order to protect native species and the complexity of their habitats, different modes were selected according to different ways of grassland utilization. By establishing different interrupting modes (including grazing, burning, mowing and no-action) and tendency-forecasting models, detailed experimental schemes were set forth (Wong and Morgan, 2007). Hou et al. (2011) indicated that fundamental research into grassland ecosystem should be enhanced, and multi-dimensional theories, knowledge, methods should be synthesized. They stressed that it is also necessary to carry out pilot demonstration projects as well as to establish the digital information network to improve the ability of supervision and forecasting.

Concerning livestock balance management, Yang (2013) emphasized the importance of reform on livestock balance management, which is transferring the command control on balancing livestock and forage base to marketable measures on adjusting the total quantity of livestock. Li and Li (2012) showed that since the heterogeneity of resource allocation is extremely significant in arid and semi-arid grasslands, the current means of balancing between livestock and forage base, which is imposed on households, are still open to discussion. They argued that grassland management on this aspect should be more concentrated on the flexible management of ecosystem rather than the threshold as the carrying capacity of its own.

With regard to the sustainable development of grasslands, domestic

research on sustainable ability is still at the starting stage since no consensus have been reached on its concepts, evaluation theory and methodology. On the theoretical aspect, Xu (2009) classified two groups of factors with impact on grassland sustainable development: natural and social factors. Deng (2004) regarded the sustainable development of grasslands as the balance state among the sustainability of ecology, economy and society. Qin (2004) revealed that grassland degradation is not only related to irrational utilization, but also the in-depth reasons behind economic system, population pressure, and bioecology. Wang (2006) analyzed theoretical and practical problems concerning the continuous utilization of grasslands, including degradation scale, uncertainties on degradation causes and its weightings, load capacity and its forecasting theories. On the evaluation method, Ha (2006) combined field surveys with qualitative and quantitative analysis to study the problems arising from grassland sustainable development. Liu et al. (2009) classified grassland resources using several indexes, including grassland productivity, value of ecosystem services, and resource classifications. Jie and Ta (2013) examined the sustainable abilities of Yili Grassland, Xilingol Grassland, Nagqu Frigid Grassland and Qilian Mountain Grassland through the method of three capacities (economic capacity, resource ecological capacity, administrative capacity). They believed that only when the three capacities reach a symbiosis of interests and a balanced development, can the grasslands be endowed with a stronger sustainability. Regarding the heterogeneity on grassland resource allocation and differentiation of regional economic development, Liu et al. (2011) selected 12 functional indexes with high weightings to evaluate the economic values of ecosystem services which would be adapted to the grassland.

As to community-based management, Zhang (2013) suggested that the current management on natural resources is more administrative-unit-based, which ignores the resource-utilization dimension of different communities. In this way, Zhang argued, the policy execution would hardly reach expecting results. One community called Gazha in Inner Mongolia was taken as an example. It is revealed that community-based grassland management should use the hierarchical analysis to set rules on occupation, provision, supervision and implementation. Setting up an administrative hierarchy on communities would ensure the coherence of local climate conditions and husbandry demands. Wang (2009) pointed out the rigid policy measures and villager-exclusion were the main causes of grassland degradation, which were also the main reasons behind the situation where current protection policies were hard to achieve their expecting results. It is stressed that efforts should be made to encourage the participation of villagers and collective actions during the policy-making and conducting.

Concerning the administrative investment, Meng et al. (2006) systematically analyzed and discussed the investment problems of the northern grasslands, including insufficient economic investment, imperfect investment structures, low profitability, insignificant income-raising, imperfect investment mechanism and the lack of sufficient policies on issues concerning herdsmen, husbandry and pasturing areas.

As to the scientific and technological (S&T) development, Hou et al. (2004) utilized the analytic hierarchy process (AHP) to evaluate the key points of S&T development of western grasslands. They suggested that grassland degradation, regulation mechanism, interconnection of ecosystems, and grazing management should be the key points to consider.

From the perspective of grassland law system, Feng and Li (2013) studied the state and local regulations on grassland management, suggesting that the formulation of law system should be in line with social development. It is also necessary, according to the researchers, to eliminate the bias during the implementation. Wan and Liu (2010) assessed the values, techniques and impacts of articles in the *Law of the People's Republic of China on Grasslands* (edited in 2002) (abbr. *The Grassland Law*). The lack of justice values, irrational logical structures, imperfect criminal regulations and laws, lack of commitment on criminal responsibilities, larger span on penalty amount are the main problems of the current laws and regulations.

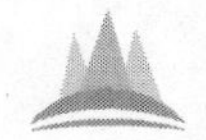

2.4 Practice of Regional Grassland Management

2.4.1 Case study on regional grassland management in foreign countries

1. Ranch Leasing Program, Australia

A contract for ranch leasing is usually signed jointly by a government and a tenant (farmer). This contract specifies the leasing interval (25-50 years) and leasing regions for grazing (mostly for cattle and/or sheep). In Queensland, ranch leasing is for both grazing and vegetation. In the southern part, western part and northern part of Australia, legislatures take into consideration more functions and public services such as gardening and agricultural tourism.

Table 2-5 Various legislative purposes of ranch leasing

Area	Legislation	Legislative Purposes
New South Wales	*Western Land Act, 1901*	Leasing should only be used for the purpose of animal raising
Queensland	*The Land Act, 1994*	During the leasing duration, animal raising is only for agricultural stocking
South Australia	*Land Management and Protection of Ranch, 1989*	The ranch leasee has no rights to use land for other purposes except for the permission of Pastoral Council. The purpose of grazing means inventory and other auxiliary purposes
West Australia	*Land Management Act, 1997*	Ranch should not be used without the permission, except for grazing. The grazing purpose means the commercial grazing with the permission, permission-related land use on agriculture and gardening, as well as other auxiliary activities
Northern Territory	*Pasture Land Law, 1992*	Leasing should only be used for the purpose of animal raising. It means the sustainable commercial or agricultural grazing or other related activities, including productions and pastoral tourism
New Zealand	*Royal Ranch Land Law, 1948*	For the leasing duration, the general requirement is for grazing purpose

Data source: Song (2006)

Justice departments specify detailed requirements, including assuming corresponding responsibilities and following the rules of land supervision. Usually, legislative departments would make a clear statement over the comprehensive obligations and then negotiate between delegates of governments and individuals or private companies over some special assets. Tenants during the leasing should obey the following requirements: ① individuals or private companies that run the business should see to it that the land utilization is in line with the leasing programs; ② they should prevent land degradation; ③ when the financial conditions are permitted, efforts should be made to enhance the conditions and utilization of land.

2. Successful experience from funded projects of Asian Development Bank

Asian Development Bank summarized the similar characteristics of its successful projects. Although not all of them were aimed at grassland improvement, they should have the dual objectives: raising income and reducing land degradation. Therefore, some of their experience would be of great significance to the implementation of other projects, which are summarized as follows.

(1) Most of the policies for the projects were made after public inquiries. The policies took the current problems of rural areas into consideration, and their commitments were devoted to improving the living standards of local people and protecting the regional environment.

(2) All the funded projects were based upon the sensible and practical evaluation of the capacity of natural resources.

(3) All the funded projects had dual objectives—raising income and protecting and restoring ecological environment.

(4) Most of the exploited activities were planned and designed by the public, and if necessary, the local sectors provided technical support.

(5) All members of the rural community, including women and children, participated in the procedures of project design, implementation and supervision.

(6) Governments at all levels and all the members of the targeted groups had a clear picture of project budgets.

(7) Funds or tangible assets were directly assigned to the targeted groups. Capitals are put in the accounts of local bank directly.

(8) The schedule and arrangement of project implementation were

somehow flexible, which was organized according to the resource endowment and labor force of the targeted groups.

(9) Technical supports were provided by the local sectors or other cooperative divisions; if necessary, technical supports were organized and managed by people who are benefited from this project or the techniques.

(10) Training courses on techniques were not disposable, and it would be better if it was organized upon further education, such as field schools for farmers. The frequencies of the training courses depended upon the seasons and educational levels of the local labor force.

(11) Related research showed that the labor demand for women of the projects is mostly negative, which should be improved through special efforts, therefore, it is of great importance to entitle women to equal training rights.

(12) In the long run, school-children-centered educational program is of profound significance since they might be future supervisors or administrators of the local environment.

(13) The public utilized the simple indexes to supervise and control the procedures. Those indexes would be synthesized to the regional and national database. The key index would be designated and included into the programs. All the stakeholders comprehended the measurement and reporting on short-run and long-run success.

(14) The key to the supervision is focusing upon the effects or results (particularly for the improvement of living standard and restoration of ecological environment), rather than investment.

2.4.2 Case study on regional grassland management in China

Yang et al. (2011) pointed out that there are three conflicts between the construction and management of natural grassland reserves in Xinjiang: ① the lack of capital investment; ② the conflict between the development of surrounding communities and the construction of natural reserves; ③ administration authorities and resource types are not clearly classified. It is necessary to establish a long-term investment system and enhance the S&T supporting ability.

In line with overgrazing and overgrazing-induced degradation theories, and various theoretical studies to address the degradation, Da and Na (2010) analyzed humanistic factors causing grassland ecological deterioration in Inner Mongolia. They proposed an integral theory concerning both restoration and development of pastoral areas. When talking about the development of pasture areas in Inner Mongolia, the researchers believed, full consideration should be given to the time-honored animal husbandry industrial layout in the Mongolia Plateau, contains nomadic culture, the layout is spontaneously formed as a result of comprehensive elements under natural economic conditions over more than a thousand years. The nomadic culture, which stresses the reasonable utilization of grassland resources, would be of great practical significance for the present grassland administration.

By analyzing the potential of carbon sink and development pathway of Inner Mongolia, Song and Tang (2012) suggested that solutions might be found at the international and domestic levels. Domestically speaking, it is necessary to improve local regulations so as to promote carbon sink increase, its evaluation and trading legislation. Internationally, based on "Clean Development Mechanism" , efforts should be made to establish an institution for grassland carbon sink and to enhance the eco-compensation mechanism through international cooperation.

Pu and Guo (2012) emphasized that traditional productive mode is the main cause of inverse evolution of grassland ecosystems in Gannan (i.e. a severe degradation on vegetation, desertification, salinization and a sharp cut-off on biodiversity). As the current productive capacity of the grasslands could satisfy the development needs of animal husbandry in the region, it is of great necessity to transform the current productive mode to ecological and industrial ones.

Wang et al. (2013) conducted a survey on the recognition of administrative policies among herdsmen in arid and semi-arid areas of the mid-western part of Gansu. Results showed that policy recognition is highly correlated to herdsmen's

educational levels, household incomes and migrant working experience. While publicizing agricultural techniques, enhancing infrastructures and specifying property rights on grassland protection, it is important to pay more attention to the characteristics of herdsmen's recognition, and then design implementation plans accepted by herdsmen.

Cai et al. (2013) carried out a field survey in the autonomous county of Zhuaxixiulong in Gansu Province. Participative evaluation was applied by collecting the data on traditional grazing modes, recognition on climate change and localized countermeasures. The current situation of public participation into policy-making and scientific research was analyzed. Policy recommendations were proposed on the basis of the survey results: localized recognition and experience from herdsmen should be fully acquired to achieve the sustainable development of alpine grasslands in the Qinghai-Tibet Plateau. This would enhance the capacity to address climate change in the field of animal husbandry and social development.

Feng et al. (2013) utilized the data from ETM and GIS to evaluate the significance of the 3 stages of eco-functional zones in Hulunbuir Pasture Land (in the years of 2000-2003, 2004-2007 and 2008-2011), especially for its sand fixation. They analyzed the significance of single factor, their total effects, their dynamic alteration and spatial distribution. The results revealed that vegetation coverage is the fatal factor producing strong impact on sand fixation. From 2000 to 2011, significance of sand fixation in eco-functional zones shows an increasing tendency, followed by a decreasing trend. Regarding its spatial distribution, significance of sand fixation occurred a gradually ascent from the east to the west. The significance became pretty high in the desert areas, but fairly low in the temperate meadow areas and temperate grassland areas.

In view of the current situation and eco-compensation efficiency of Xilingol Grasslands, Cao (2013) deemed the main problems of grassland eco-compensation are the unstable policies, incomplete coverage of compensated entities, lower standard levels, monotonous compensating means and unreasonable timing. The author affirmed that compensation should follow the rules of equity, integrity, endogeneity and regularity to make compensation more diversified and legitimate. It would be efficient to transform the eco-compensation from exogenous mode to endogenous mode, that is, ecological-oriented compensation mode to the cooperative development mode.

Hong (2012) conducted a face-to-face survey in six representative counties in Qinghai Province, Xingjiang Uygur Autonomous Region and Inner Mongolia Autonomous Region respectively. Through quantitative and qualitative data analysis, the implementation and efficiency of grassland management measures were primitively assessed. Social, economic and cognitive factors that may have the potential influence on the willingness-to-accept of ecological compensation were correlated. A regression analysis was also called for over the Tobit model to assess the potential factors that may influence the implementation of grassland protection.

2.5 Discussion

An analysis of the above literature shows that China's western grasslands are facing the following problems.

1. Development of pastoral resources

Generally speaking, regarding, there is a lack of sensible awareness of resource protection in the ecological management of grasslands. Local grazing and husbandry infrastructures are insufficient and projects related to "green for grain" have not received significant support. Financial supporting mechanism has not been feasibly established; supporting measures for eco-migration are insufficient. S&T services need improvement so that intensive grazing would be possible. Mechanisms behind disaster-resistance should be improved in order to protect the achievements of eco-environment control. Funding is not sufficient in the following aspects: grass disease and pest control, technical training, forage

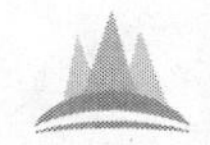

supply and resource supervision.

2. Ecological compensation and eco-migrants

Generally speaking, the compensation standards are pretty low, and the existing way of compensation payment would cause the problem of fairness. Nomadic herdsmen are not included in the eco-migration program. Eco-migrants would have difficulties in finding jobs in the new surroundings as it is hard to set up the follow-up industries in a short run. Also, it would be a hard task for herdsmen to have finance support for the grassland restoration and self-survival. It is urgent for the governments to help the eco-migrants find a way out of these difficulties.

Moreover, there is a lack of research into the negative externality resulting from grassland ecological restoration and construction. It is necessary to evaluate the negative effect that would weaken the expected results of the policies and measures, especially those on social economic development and livelihood. Regarding eco-migration, major problems to be tacked include the capacity of psychological endurance, protection of traditional culture, micro-construction of settlement models, ethnic problems and community work during the urbanization, settlement policies and their implementation impact.

3. Institutional issues

Increasing conflicts about land property call for the improvement of the property right institutions. It is difficult to conduct grassland supervision, the stability of related policies is low, and the grassland implementation capacity is weak. Meanwhile, policy-making is not highly correlated with the social security system. Taking ecological compensation as an example, herdsmen and farmers are not certain about the continuity of the current policies and potential risks they will face, which may be caused by information asymmetry.

4. Lessons from previous practice

Measures should be carried out to ensure the public participation and supervision during the whole process of policy-making and implementation; responsibilities and duties should be specified among administrators at all levels and risk-undertakers; the allocation and effect of capital investment should also be specified.

Chapter 3
An Analysis of Grassland Management Policies

3.1 Literature Review over Regulations and Laws at the National Level

Controlling the continuous ecological deterioration of grasslands is a key and urgent issue for ensuring the national ecological security and the sustainable socioeconomic development of the country. Previous research revealed that in the mid 1980s, the state investment to the grasslands in northern China, including Inner Mongolia, Xingjiang, Qinghai and Tibet, was about 4.7 billion yuan, about 15 yuan per hectare on average. However, the average annual investment over the 37 years since 1949 was only 0.3 to 0.45 yuan per hectare. The figure reached 0.75 yuan in the 1990s. The National Program of Western Development in the late 1990s promoted the ecological construction of grasslands, and the financial investment to the grasslands peaked in this period. On December 16th, 2002, 11 provinces in the western part of China officially launched the program of returning grazing lands to grasslands, showing that the government places a high priority on grassland ecological protection.

Generally speaking, China has paid more attention to grassland ecological restoration and issued a series of policies to give impetus to grassland sustainable development. The following is a review of major grassland management measures in light of policies, regulations and laws at the national level in the 21st century.

3.1.1 Policies at the national level

Table 3-1 shows the important points and measures in the Central Document No.1 from the year 2005 to 2014. Main characteristics of the previous policies and alterations in the recent decades are summarized into the following aspects.

(1) Policy focus has shifted from protection to reward-compensation. That is, the earlier stage of policy-making is concentrated upon grassland protection and ecological efficiency while the later stage of policy-making concerns the interests of herdsmen and the compensation measures become more diversified.

(2) Policy orientation has changed from being monotonous to being diversified and multi-hierarchical. In view of returning grazing to grasslands, the policies give more importance to issues concerning technical application, institution and supervision.

(3) More attention has been paid to the awareness of grassland protection in terms of implementation, means and intensity of execution.

Table 3-1 A summary of key views and measures in Central Documents No.1 (2005-2014)

Year	Grassland protection policies
2005	Further strengthening grassland construction and protection, accelerating the project of returning grazing land to no grazing grassland, and promoting hydrological construction in production areas of grazing-banning and grazing-off
2006	Encouraging the project of returning grazing land to no grazing grassland and the comprehensive development of mountainous areas; Introducing and improving ecological compensation mechanism
2007	Improving the funding mechanism for forest ecological compensation, and exploring the ecological compensation mechanism for grasslands; Accelerating the execution of the project of returning grazing land to no grazing grassland
2008	Establishing and improving the forest, grassland and water and soil conservation compensation system, and expanding financing channels to raise funds to enhance the ecological function of grasslands; Implementing the livestock-forage balance system, and promoting the development of pastoral areas, water resources, and construction of artificial grasslands

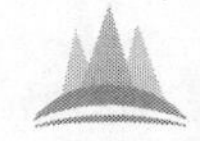

Continued

Year	Grassland protection policies
2009	Expanding the scope of the project of returning grazing to no grazing grassland, and strengthening the construction of artificial grasslands and irrigation projects; Raising the compensation standard for ecological benefits from the central government, and starting the pilot project of compensation for the ecological benefits of grassland, wetland and water and soil conservation
2010	Making more efforts on raising funds for the ecological benefits of forest, grassland, soil and water conservation; Effectively strengthening grassland ecological protection, further enforcing the project of returning grazing to no grazing grassland, extending its implementation period and appropriately raising its subsidy standards; Implementing the livestock-forage balance system, continuing the grazing-banning, grazing-off and rotational grazing policy, developing captive breeding, doing a good job of artificial forage grassland and pasturing area water conservancy construction
2011	Strengthening water conservancy construction, giving first priority to people's livelihood; Making more efforts on soil and water conservation and ecological protection of water resources
2012	Expanding the the scope of grazing withdrawal project and supporting the development of grassland fences, forage bases and barn circles, and encouraging the improvement of severe degradation grasslands; Strengthening the supervision of pastoral and semi-pastoral areas
2013	Exploring the introduction of a strict access system for industrial and commercial enterprises to lease woodlands or grasslands from farmers; Accelerating the pastoral grassland contracting practice, and starting pilot projects for pastoral grassland contract management rights and registration; Continuing the reward-compensation policy on grassland ecological protection
2014	Continuing the projects of returning farmland to forest and grassland in steep-slope farmland, serious desertification farmland and important water resources areas; Strengthening the implementation of the projects of returning farmland to forest and grassland, and starting the projects of "southern grassland development and utilization" and " nature reserve development on grasslands"

3.1.2 Regulations and laws at the national level

Table 3-2 shows that regulations and laws regarding grassland management have become more detailed and specified. Since the implementation of *the Grassland Law of the People's Republic of China*, more supplementary details and directives have been issued continuously.

Main measures to promote grassland sustainable development could be seen in the following documents: Opinions of the State Council on strengthening the protection and construction of grassland, National subsidy policies on grassland ecological protection and Opinions on improving the policy of returning grazing to grassland. Their core points include the following:

1) Introducing a primary grassland protection system;

2) Enforcing a livestock-forage balance system;

3) Introducing such systems as rotational grazing, grazing-off and grazing-banning, and offering incentives or subsidies for herdsmen participating in the projects of grazing-banning and livestock-forage balancing.

Table 3-2 Relevant laws and regulations on grassland protection (2002-2011)

Year	Laws/Regulations	Major points
2002	*Grassland Law of the People's Republic of China* (revised)	The grasslands are owned by the State, with the exception of the grasslands owned by collectives as provided for by law. The State-owned grasslands may, in accordance with law, be assigned for use to the units under the ownership by the whole people and to collective economic organizations. The right to contractual management of grasslands is protected by law, and it may be transferred in accordance with the principles of voluntariness and compensation
2002	*Opinions of the State Council on Strengthening the Protection and Construction of Grassland*	Establishing systems of basic grassland protection, livestock-forage balancing, rotational grazing, grazing-off and grazing-banning
2003	*Opinions on Improving the Policy of Returning Grazing to Grassland*	Establishing systems of basic grassland protection; livestock-forage balancing, rotational grazing, grazing-off and grazing-banning. Improving their implementation plans and launching related projects; promoting household contract system and specifying the rights and obligations; strengthening supervision and management to balance livestock and forage; strengthening the support on science and technology and production mode transformation
2005	*Animal Husbandry Law of the People's Republic of China*	The State supports the development of animal husbandry, and gives full play to the role of animal husbandry in agriculture and rural economy development and farmers' income improvement. The State supports the development of animal husbandry in ethnic minority areas and poverty-stricken areas. Grasslands should be protected and rationally used, and the conditions for the production of animal husbandry should be improved
2006	*Administrative Measures on Examination and Approval on the Occupation of Grasslands*	Strengthening the supervision and management of grassland appropriation, protecting grassland resources and the environment, and safeguarding the legitimate rights and interests of farmers and herdsmen
2010	*National Subsidy Policies on Grassland Ecological Protection*	Formulating the specific standards for subsiding and rewarding grassland ecological protection projects such as grazing banning, livestock-forage balancing and production
2011	*Opinions on Promoting Sound and Rapid Development of Pastoral Areas*	It is pointed out the meaning, principles and general requirements on promoting sound and rapid development of pastoral areas and more opinions were attached, including: strengthening ecological protection and construction of grassland, and improving the capacity for sustainable development; accelerating the transformation of the mode of development, and actively developing modern grassland animal husbandry; promoting economic development in pastoral areas, raising the herdsmen income, and broading their employment channels; vigorously developing the public utilities, effectively protecting and improving the livelihood of the people; strengthening the organization and leadership of pastoral work
2011	*Opinions on Improving the Policy of Returning Grazing to Grassland*	Imposing requirements on improving the policy of returning grazing to grasslands, including: reasonable layout of grassland fence; construction of feeding stalls and artificial forage bases; increasing the proportion of central investment subsidies and standards; changing feed grain subsidy into grassland ecological protection subsidies and awards

3.2 Literature Review on Regulations and Policies of Grassland Management in the Heihe River Basin

3.2.1 Grassland management policies and regulations in Gansu Province

The following literature review on regulations and policies of grassland management mainly focuses on Gansu Province since prefecture-level cities in this administrative unit occupies the main areas of the Heihe River Basin.

1. Grassland Regulations in Gansu Province

Grassland Regulations in Gansu Province was approved on December 1st, 2006 and put in force on March 1st, 2007. Its main contents include general principles of the implementation, including the scope of implementation and execution sectors. In the grassland planning and construction part, it illustrates the implementation and execution of grassland planning, field survey, technical application, importance and means of disaster resistance. Ownership property and contracting management have also been introduced. Protection and utilization of grasslands are explained in details as well as the cognition of legal liability due to the violation of regulations. The regulations has also specified the management policies on grassland sustainable development, including the following.

(1) The quantity of livestock raised by grassland users or contract operators should not exceed the approved carrying capacity. The total amount of accessible forage should be dynamically balanced with the demanding amount of livestock.

(2) Grassland users and contract operators should not rely on the traditional productive mode, and measures that can enhance the comprehensive productive capacity of grassland should be adopted, including grazing-banning, rotational grazing, grazing-off, drylot feeding and captive breeding.

(3) Prohibition on digging turfs and peat to prevent the vegetation damage, grassland desertification, water loss and soil erosion.

(4) Practices and measures on grazing-off and grazing-banning should

follow the regulations that are implemented and issued by the State Council and the provincial governments.

(5) Fees for vegetation recovery should be charged according to the regulations that are implemented and issued by the State Council and the provincial governments.

2. Administrative Measures on Grazing-Banning in Gansu Province

Administrative Measures on Grazing-Banning in Gansu Province was approved on November 20th, 2012 and put in force on January 1st, 2013. Main measures include the following.

(1) The practice of grazing-banning should be introduced in areas suffering from severe degradation, desertification and salinization, as well as ecologically vulnerable zones and important water conservation zones.

(2) Governmental grassland management and supervision administrations above the county level should be responsible for enforcing grazing-banning, and their expenditures should be included into the fiscal budget at the corresponding level.

(3) Grazing-banning zones should be defined and designated by the grassland administrations strictly according to ecological monitoring.

(4) Grassland grazing-banning should be fit into the responsibility system of objective management of governments and authorities at all levels. Letters of responsibilities should be issued and signed at all levels.

(5) Posts of grassland attendants should be set up in the villages and communities where eco-protection reward-subsidy compensation system has put in force. Their responsibilities and duties should be specified.

(6) For local governments and administrative authorities above county level and township governments, systems of inspection, monitoring, reporting, subsidy and publicity concerning grazing-banning should be set up and improved.

3. Grassland ecological reward and subsidy in Gansu Province

Provincial conference on implementing incentive polices for grassland

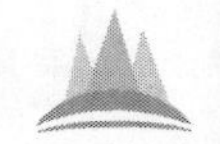

ecological protection in Gansu Province was held on July 31st, 2011, which specified the targets of ecological compensation. Grassland ecological compensation fund from the central government amounted to 1.14 billion yuan for 16 million hm^2 of available grasslands in the province. This amount of funds is mainly used for subsidizing herdsmen involved in grazing-banning, and superior breeds of animal husbandry and forage. Measures are summarized as follows.

(1) Compensation standard of grazing-banning was specified according to the scientific measurement and assessment, which is: 20 yuan per mu on the Qinghai-Tibet Plateau, 2.95 yuan per mu on the Loess Plateau, and 2.2 yuan per mu in western desert region. Subsidies are paid directly to households through bank accounts.

(2) Except for grazing-banning zones, according to the verification of Central Government, all the remaining 9.4 million hm^2 of available grassland in Gansu Province are classified as the balancing areas for forage and livestock, which will receive subsidies totaling 210 million yuan. The average overloading rate of livestock in this area reaches 30.6%, which should be eliminated within three years.

(3) According to the Central Government verification, the coverage of artificial grasslands in Gansu amounts to 1.47 million hm^2. The subsidies for superior breeds of forage in 2016 will amount to more than 220 million yuan, which will be used to replace remaining forage, expend the plantation coverage and implement pilots for disaster-resistance and livestock-breeding.

4. Regulations on Balancing Livestock and Forage in Gansu Province

Regulations on Balancing Livestock and Forage in Gansu Province was approved and issued on September 18th, 2012 and put in force on November 1st, 2012. Its major content is summarized as follows. Governments at all levels are responsible for publicizing the knowledge on balancing livestock and forage, and providing support and guidance for farmers and herdsmen to adopt more sensible measures on artificial grassland recovery, breeding improvement, drylot feeding and captive breeding. Cutting down the loading capacity of grasslands would be beneficial to preventing grasslands from over-grazing. When verifying their local loading capacities of livestock and forage, it is necessary for township governments and administrative authorities to consult expertise panels and listen to the opinions of grassland users and contracting operators. According to the State regulations and policies on ecological compensation system, rewards and subsidies are allocated to the farmers and herdsmen who follow the rules and take actions on balancing livestock and forage. The compensation is directly paid to each household or contracting household.

3.2.2 Grassland management policies and regulations in Inner Mongolia Autonomous Region

1. Forest and Grassland Fire Prevention Act in Inner Mongolia Autonomous Region

Forest and Grassland Fire Prevention Act in Inner Mongolia Autonomous Region was approved and issued on September 18th, 2012 and put in force on November 1st, 2012. It emphasizes fire-precaution and fire-fighting on grasslands in order to protect forest and grassland resources, showing that high priority is given to fire prevention in grassland management.

1) Fire-fighting and fire-precautionary organization

Local governments should set up headquarters on fire-fighting and precaution by joining hands with local forces of PLA and armed police. On behalf of local governments of autonomous region or counties, Forestry Administration of Daxinganling Mountains and its state forestry administration should organize and set up the fire-precautionary headquarters to take charge of the fire-precautionary work. Executive offices should be established by local governments above county levels to take charge of the routine work of headquarters.

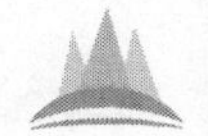

2) Precaution on forest and grassland fire

Local governments at all levels should be in charge of designating the guard units and specifying the accountability units. The responsibility system should be improved and completed by the regular inspections. Institutions or individuals are of no exceptions to implement the precautionary responsibility and should obey the supervision of the local headquarters.

3) Firefighting in forests and on grasslands

Any unit or individual has the responsibility to report the fire-related issues to local governments or headquarters on firefighting, and should take actions to fight fire immediately. Any unit or individual should facilitate the fire-reporting for free. It is forbidden to report false alarms.

4) Dealing with fire aftermath

After a fire is put off, it is urgent for the local governments to re-schedule the vegetation recovery work and put it into practice. The extermination of diseases and insect pests over stricken areas and the quarantine on human and livestock diseases in the region are essential to prevent epidemic diseases from occurrence and spreading.

2. Regulations for Grassland Management in Inner Mongolia Autonomous Region

Regulations for Grassland Management in Inner Mongolia Autonomous Region was revised and approved on November 26th, 2004, and put in force on January 1st, 2005. The main content of the regulations includes the general principles of its implementation, grassland property and its transfer, grassland planning, its protection and supervision, and the legal liabilities for grassland utilization. Policies concerning grassland sustainable development include the following.

(1) Prohibition against grassland reclamation. For the cultivated land suffering from severe soil erosion, with desertification trend and in need of ecological recovery, it is necessary to return them to grasslands in a planned and

orderly way. The grasslands already suffering from degradation, desertification and salinization should be treated within a definite time period.

(2) Balancing between livestock and forage. Local governments and administrative authorities at the county level should conduct triennial reviews on the forage-livestock balance.

(3) Operators of the State or collective-owned grasslands should sign a letter of responsibilities on forage-livestock balance according to the approved livestock carrying capacity.

(4) Areas and time intervals for grazing-banning and grazing-off should be announced by local governments at the county level. Grazing is forbidden during the grazing-banning or grazing-off period in the designated protective areas.

(5) Any construction programs on grasslands and artificial grassland projects (over small areas) that may alter or influence the surroundings or native vegetations should be in compliance with the local planning and principles of grassland protection and utilization. Local governments above county level should strengthen their supervision and inspection.

(6) It is forbidden to plant rain-fed artificial pastures with the following characteristics: annual average precipitation is lower than 250 mm; gradient is higher than 20°; properties and conditions of soil are not appropriate for plantation.

(7) It is forbidden to dig vegetation or to carry out other destructive activities in the ecological vulnerable areas or over the grasslands suffering from soil erosion, (semi-) desertification, salinization and severe degradation.

3. Provisional Regulations on Balancing Livestock and Forage in Inner Mongolia Autonomous Region

Provisional Regulations on Balancing Livestock and Forage in Inner Mongolia Autonomous Region was put in force on August 1st, 2000 in order to realize the sustainable development of local husbandry as well as the rational utilization and protection of grasslands. It gives equal importance to husbandry

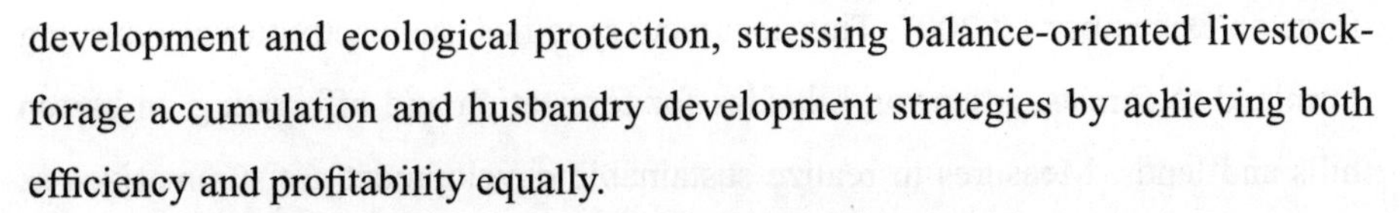

development and ecological protection, stressing balance-oriented livestock-forage accumulation and husbandry development strategies by achieving both efficiency and profitability equally.

1) Verification and approval on livestock-forage balance

Local governments and administrative authorities at the county level should conduct biennial reviews on forage-livestock balance, that is, to ratify the total storage of forage and the appropriate livestock carrying capacity. The balancing should be conducted by the grassland contracting operators or institutions that have the rights to use the grassland. For those who have no contract on operation, balancing should be conducted by the institutions who have the rights to use or who have the ownership.

2) Management on livestock-forage balance

When the quantity of livestock exceeds the carrying capacity, contracting operators should take following actions: planting and reserving forage, and increasing the supply of forage; drylot feeding and periodic grazing-off or zoning rotational grazing; optimizing the structure of husbandry and improving the rate of livestock delivered to the slaughter house.

3) Reward and punishment

Local governments at all levels should reward the grassland contracting operators, users and owners who have followed the rules on livestock-forage balance and devoted themselves to grassland protection. For those violating the rules or refusing to sign the letters of responsibilities, local governments above the county level should issue warnings and carry out punishment timely for the violating behaviors including livestock overloading and overgrazing on the grassland.

4. Administrative Measures on Returning Farmland to forests in Inner Mongolia Autonomous Region

Administrative Measures on Returning Farmland to Forests in Inner Mongolia Autonomous Region was approved on September 27th, 2007 and put in force on December 1st 2007. This document specifies such activities as returning farmland to forests, reserving hillsides for forestation and afforesting on barren hills and lands. Measures to realize sustainable development are summarized as follows.

(1) Local governments above the county level should not only take actions on returning farmland to forests, rural infrastructure and energy construction, eco-migration and follow-up industrial development, but also synthesize and reinforce the achievements of the above actions.

(2) Local governments above the county level should support the research and application of techniques for returning farmland to forests.

(3) When distributing food subsidies, local governments at the county level should take the rate of forest survival and retention rate into consideration.

(4) Local governments at all levels should hold frequent consultations with individuals and units that are responsible for returning farmland to forest and afforesting on the barren hills and lands, as well as the centralized procurement on seedlings.

(5) Financial departments should arrange special funds as local supporting funds for returning farmland to forest. It is used for preparation, administration and inspection.

(6) It is forbidden to conduct activities deteriorating the vegetation in the areas that have just been recovered through returning farmland to forest.

3.2.3 Grassland management policies and regulations in Qinghai Province

1. Measures on Implementing "The Grassland Law" in Qinghai Province

Measures on Implementing "The Grassland Law" in Qinghai Province was approved on September 28th, 2007. This document specifies the measures and regulations on grassland planning, protection, utilization and management, which

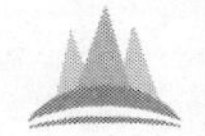

are summarized as follows.

1) Grassland ownership and obligation

Within the administrative areas of the province, the State-owned grasslands that are legally determined to be used by the State-owned units or collective economic organizations should be registered at the people's governments at or above the county level to get warrants and confirmation on grassland use rights. Any changes in grassland ownership should be registered. Besides, the duration of operation contracting shall be 50 years, and contracting operators should follow the corresponding rules and keep the ownership after returning farmland to forest within the duration.

2) Grassland planning and exploitation

Administrative authorities at all levels should meet the requirements of the upper levels to formulate the planning on grassland protection, construction and utilization within its terms of reference. If there is any necessity to adjust or modify the planning, it should be permitted and ratified by the approval authorities. Besides, local governments at or above the county level should cooperate with other related sectors to conduct dynamic monitoring on grassland basic conditions, including the amount and level of areas, vegetation composition, productive capacity, ecological condition, natural disasters and biological disasters. Monitoring results, as well as the information on early warning should be shared and mutually exchanged.

3) Grassland utilization

Implementing livestock-forage balance system and corresponding measures step by step is essential for grassland utilization. Administrative authorities at all levels should ratify and announce livestock carrying capacity as well as the breeding quantity for contracting operators and grassland users according to the detailed standards and information on basic conditions, productive capacity as well as monitoring results of grassland within its own administrative units. Measures on drylot raising and stall breeding, improvement on slaughter rates and husbandry structure should be adopted in order to rationally utilize grassland resources as well as balance the livestock and forage.

4) Grassland protection

Any institution or individual should not damage the grassland protective infrastructures or signs. Unauthorized occupation or alteration on application is also forbidden. Local governments at all levels should introduce systems on returning farmland (grazing land) to grassland, grazing-banning and grazing-off in conformity with legal provisions. It calls for vegetation protection on shrub lands, and the rational and timely utilization and protection on rare and endangered species. Besides, local governments should establish and improve the responsibility system on grassland fire prevention, and the fire-precautionary interval is from September 15th to July 15th in the following year.

5) Supervision and management

Local governments and administrative authorities at or above the county level should perform the following duties: the publicity on grassland laws, regulations and rules, supervision and inspection on execution; the punishment on the violation behaviors in conformity with legal provisions; responsibilities on contracting operation, registration, tabulation and warrant-issue on the right of use; the conciliation on ownership dispute; dynamic monitoring and supervision upon livestock-forage balance; the cooperation and coordination on fire-prevention; other supervision-related management.

2. *Regulations on Grassland Ecological Reward-Subsidy Compensation Mechanism in Qinghai Province (trial scheme)*

The trial scheme of *Regulations on Grassland Ecological Reward-Subsidy Compensation Mechanism in Qinghai Province* was issued on September 28th, 2011, and the compensation mechanism on ecological protection reward and subsidy was put on the agenda shortly afterwards.

(1) In order to protect the natural pasture, subsidy policies and standards on grazing-banning are implemented and formulated. Subsidy standards and total amount of funding are designated according to the state measurement standards.

Funds are distributed to each household in view of the real situation of each administrative area.

(2) Subsidy policies on balancing livestock and forage are formulated and implemented by ratifying the overloading amount of livestock, planning on reducing livestock, and establishing and implementing system on balancing livestock and forage. Financial sectors and agriculture and husbandry departments at all levels should subsidize the herdsmen who plan to reduce livestock with the measurement standard of 1.5 yuan per mu per annum.

(3) Subsidy policies on productive activities for herdsmen should be put into practice. Subsidies should be distributed according to the practical needs on superior breeds. Subsidy standards are 2000 yuan for each yak, 800 yuan for each Tibetan sheep, and 800 yuan for each Cashmere goat. Besides, subsidy on artificial seeds in pastoral regions and farming-pastoral regions should be distributed as well.

(4) Performance appraisal and reward system would be essential for reinforcing the local subsidy-reward policies as well as the protection achievements. This is also the important basis to ensure the rational management and utilization of supporting subsidy funds and to carry out the performance appraisal. Local governments that have achieved remarkable performance and success should be rewarded, and those that have not fulfilled the tasks should be punished and criticized.

Chapter 4

Spatial Analysis of Main Habitat Factors in the Heihe River Basin

The Heihe River Basin, the second largest inland river catchment in China, bears all natural characteristics of an inland river basin in an arid region. The Heihe River originates from two tributaries in the southern Qilian Mountains. One tributary is called the EBo River (also known as the Babao River) and flows about 80 km, and the other, which is to its west, is named the Yeniugou River and runs from the west to the east totaling 190 km. The two tributaries join in Huangzang Temple to form the Ganzhou River flowing north. The Ganzhou River runs through 90 km, reaching the Yingluo Gorge. Crossing the gorge, it flows in the middle Hexi Corridor, and is named the Heihe River. The section from headstreams to the Yingluo Gorge belongs to the upper reaches of the Heihe River. The Heihe River runs 185 km across the middle Hexi Corridor, and reaches the Zhengyi Gorge, then goes into the northern Alxa Highland. The section between Yingluo Gorge and Zhengyi Gorge is the middle reaches of the Heihe River. After the Zhengyi Gorge, the river is divided into two branches, that is, the East River and the West River. The two branches end up in two terminal lakes, respectively. The East River drains into the East Juyan Lake and the West River goes into the West Juyan Lake. The section between the Zhengyi Gorge and two Juyan lakes is the lower reaches of the Heihe River. The northern part of the basin borders on Mongolia, the eastern part is connected with the Shiyanghe River Basin, and the western part is adjacent to the Shule River basin. Administratively, the basin includes three provinces (Qinghai, Gansu and Inner Mongolia). The upper reaches consist of Qilian County of Qinghai Province and Sunan County of Gan-

su Province, the middle reaches cover Shandan, Mingle, Zhangye, Linze, Gaotai, Jiuquan, which belong to Gansu Province, and the lower reaches contain Jinta County of Gansu Province and Ejina Banner County of Inner Mongolia.

The Heihe River Basin consists of three major geomorphologic units, that is, the southern Qilian Mountains, the middle Hexi Corridor and the northern Alxa Highland（Wang, 2013）. The south Qilian Mountains with an elevation from 2400 m to 5520 m belong to the upper reaches. They are characterized by higher terrain. The middle Hexi Corridor with an elevation from 1100 m to 2200 m is the middle reaches. The northern Alxa Highland is the lower reaches with an altitude of less than 1000 m in most of the area except for the western Mazong mountains (Figure 4-1). Because the basin is far from oceans, it is mainly controlled by the continental climate characterized by sparse rainfall, windy, strong sunlight, and large daily temperature difference (Zhao et al., 2005). At the same time, the climate is coupled by the Qinghai-Tibet Plateau climate. The upper reaches express the features of mountain climate. Therefore, the difference of climate conditions in the basin is significant, which resulted in obvious vegetation and soil variation.

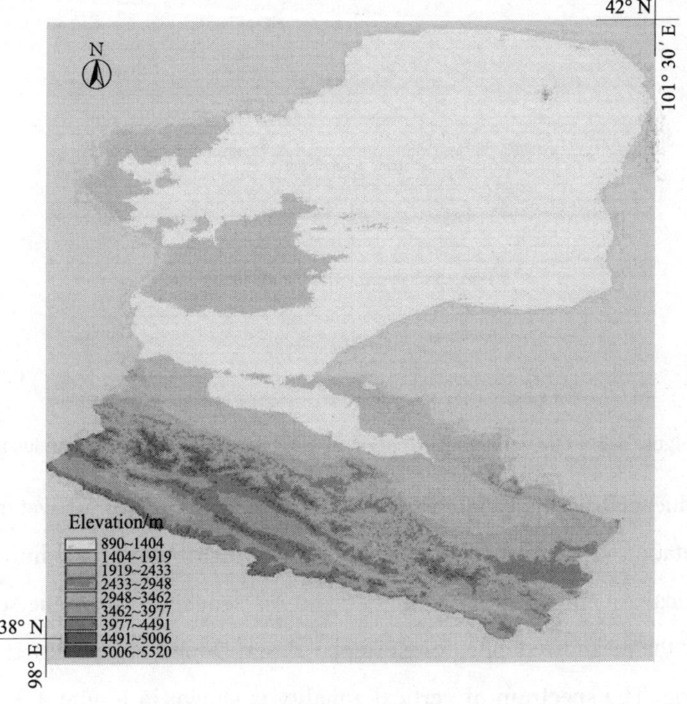

Figure 4-1　DEM of the Heihe River Basin

There are 21 types of soil in the basin, of which, 16 types belong to the zonal soil, and 5 the non-zonal soil. In the upper reaches, the main soil types include cold desert soil, alpine meadow soil, alpine shrub meadow soil, alpine steppe soil, subalpine meadow soil, subalpine steppe soil, grey cinnamon soil, mountain chernozem soil, mountain chestnut soil, and mountain sierozem. Figure 4-2 shows its vertical soil spectrum in the upper reaches. The distribution area of grey brown desert soil and grey desert soil is located in the middle and lower reaches. In addition to the zonal soil mentioned above, there are non-zonal soils, such as irrigation silting soil (oasis irrigation farming soil), saline soil, fluvo aquic soil, potential fertility soil (swamp soil) and sandy soil. Within the territory of the downstream (Ejina Banner), gray brown desert soil is mainly zonal soil. Affected by underground water level, climate and vegetation, the non-zonal soils appear, including sulfate salinized fluvo aquic soil, forest-shrub meadow soil and saline forest-shrub meadow soil, alkali soil, meadow saline soil, aeolian sandy soil and chap land soil.

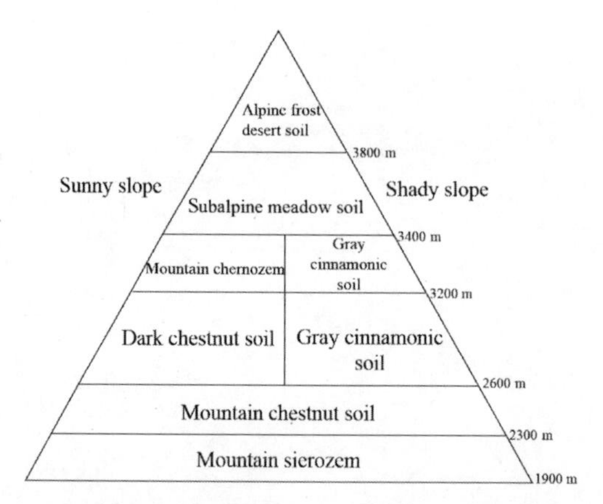

Figure 4-2 The vertical distribution of soil types in Qilian Mountainous area

Influenced by many factors, especially the combination of water and heat, the vegetation in the basin shows obvious zonal distribution including horizontal and vertical zonality. The horizontal zonality is sequenced from the south to the north as: the meadow steppe, forest steppe, desert steppe, steppe desert and desert vegetation. The spectrum of vertical zonality is shown in Figure 4-3: the alpine

mat vegetation is in the altitude of 4000-4500 m; the alpine meadow vegetation is lain in 3800-4000m; the alpine shrub meadow is located in 3200-3800 m altitude; the montane forest steppe is generally distributed in the altitude of 2800-3200 m; the area from 2300 m to 2800 m altitude is dominated by the mountain dry steppe zone; the steppe desert zone exists in the zone of 2300-2000 m (Wang et al., 2001). Zonal vegetation of downstream in the basin is desert vegetation, being made of small temperate shrub and semi shrub. There are artificial forests at piedmont alluvial fan and the alluvial plain of the middle reach, showing artificial vegetation landscape. In the riparian of downstream, and around two terminal lakes, there exists a natural oasis landscape at desert and Gobi background.

Because of the change of natural conditions, there are many kinds of ecosystems such as mountainous ecosystem, oasis ecosystem, desert ecosystem and so on, which lay a foundation for the diversity of grassland types. Regional economy depends on grassland and irrigation conditions including the mountainous pastoral areas, corridor oasis agri-pastoral areas and north plateau pastoral areas. By collecting the data of observation and survey in the Heihe River Basin, the natural characteristics of the basin are analyzed, providing important prerequisites for understanding the distribution of grassland.

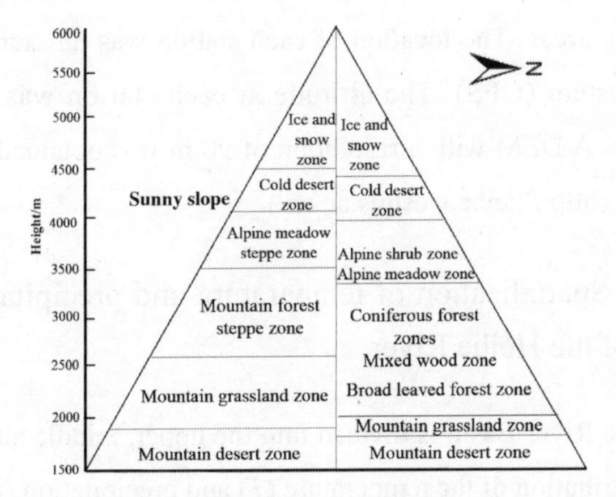

Figure 4-3 The vertical distribution of vegetation types in Qilian Mountainous area

Among many significant habitat factors, temperature and precipitation

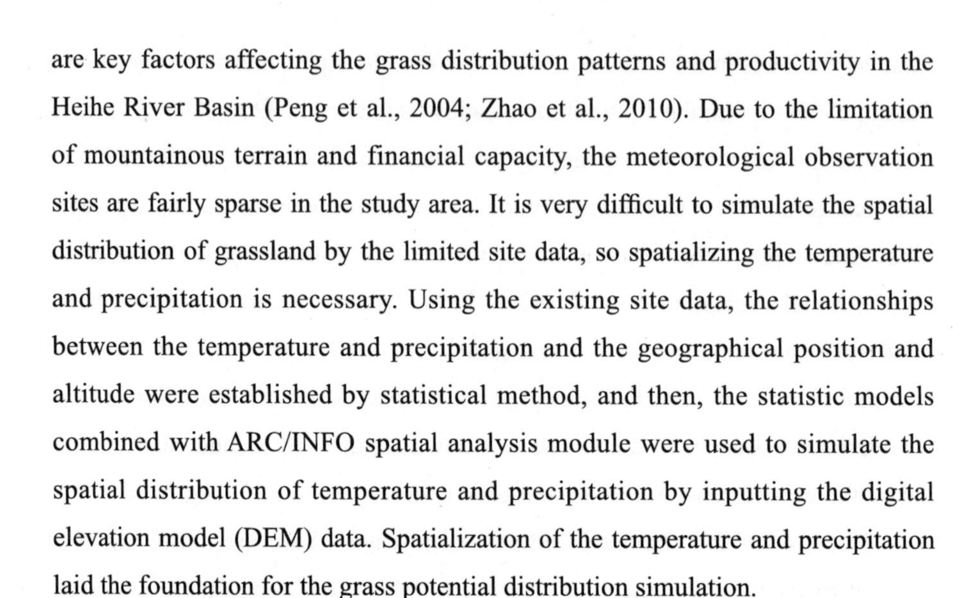

are key factors affecting the grass distribution patterns and productivity in the Heihe River Basin (Peng et al., 2004; Zhao et al., 2010). Due to the limitation of mountainous terrain and financial capacity, the meteorological observation sites are fairly sparse in the study area. It is very difficult to simulate the spatial distribution of grassland by the limited site data, so spatializing the temperature and precipitation is necessary. Using the existing site data, the relationships between the temperature and precipitation and the geographical position and altitude were established by statistical method, and then, the statistic models combined with ARC/INFO spatial analysis module were used to simulate the spatial distribution of temperature and precipitation by inputting the digital elevation model (DEM) data. Spatialization of the temperature and precipitation laid the foundation for the grass potential distribution simulation.

4.1 Materials and Methods

4.1.1 Data collection

The meteorological data (from 1960 to 2014) were obtained from 30 rainfall stations and 22 meteorological stations located within the study and the surrounding areas. The location of each station was measured with a global positioning system (GPS). The altitude at each station was recorded by an elevation meter. A DEM with a resolution of 90 m was obtained from the digital Heihe website (http://heihe.westgis.ac.cn/).

4.1.2 Spatialization of temperature and precipitation in the upper reaches of the Heihe River

The Heihe River Basin is divided into the upper, middle and lower reaches. The spatial distribution of the temperature (T) and precipitation (P) was simulated in the three parts, respectively, according to derived available data and terrain condition. As the upper reaches are occupied by high elevation mountains, the

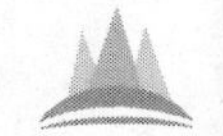

temperature and precipitation are affected by altitude and express the vertical distribution pattern, that is, the temperature decreases with the elevation increase, the precipitation increases with elevation increase. Besides, geographical location, differently influenced by oceanic air masses, is an important factor to affect the precipitation and temperature. Therefore, altitude, longitude and latitude are selected as independent variables in the distribution models of precipitation and temperature (Zhao et al., 2005). SPSS software was used to analyze the correlation of the annual mean temperature and precipitation data in the upper reaches of the Heihe River with the longitude, latitude and altitude, and the relationship is expressed as follows.

$$P = a + bH + cY + dX \tag{4-1}$$

$$T = a + bH + cY + dX \tag{4-2}$$

where X is the longitude in degree; Y the latitude in degree; H is the altitude in meter; a, b, c and d are the regression coefficients.

4.1.3 Spatialization of temperature and precipitation in the middle reaches of the Heihe River

In the middle reaches of the Heihe River, the land was flat. Meteorological observation stations in the area are relatively more than other two reaches. The spatial distribution of temperature and precipitation in the reach was simulated by interpolation analysis (Zhao et al., 2007). By comparing many kinds of interpolation models, we found that ordinary kriging (OK) interpolation model had the higher accuracy. Therefore, OK was selected to simulate the distribution of temperature and precipitation in the middle reach.

4.1.4 Spatialization of temperature and precipitation in the lower reaches of the Heihe River

The area of the lower reach of the Heihe River accounts for nearly 70%

of the entire basin, but only 6 meteorological stations are distributed in the area. Because of scarcity of stations, there was no statistical significance to build regression models, and it is hard to use the spatial interpolation methods. Thus, the analysis of the temperature and precipitation in the region was made according to the spatial distribution of the meteorological stations. We obtained the differences of temperature and precipitation from north to south in the reach, which do not affect the grassland distribution because the extreme arid resulting from the combination of *P* and *T*.

1. Simulating the potential distribution of grassland

Potential vegetation implies that over long periods and in the absence of disturbance, similar types of plant communities (PVTs) will develop on similar sites (Franklin, 1995). That is, the potential vegetation, as a final state of succession which achieves the balance with its site, is the most stable and mature climax vegetation type at the site without human interference. Knowing which PVT will develop is useful because it indicates ecological site potential and helps predict ecosystem response to disturbance processes such as wildfire and invasive species (Zhao et al., 2007). Potential vegetation offers insights about inherent productivity and other vegetation-site relationships, is valuable for projecting plant succession pathways, and helps put the existing vegetation patterns of an area into an ecological context. Potential vegetation can be thought of as the algebraic sum of all environmental factors influencing the flora of an area. The physical factors of a plant's environment interact to form a temperature and moisture regime influenced primarily by gradients of elevation, slope steepness and configuration, aspect, and geologic parent materials and their resulting soils. These site factors interact by complementing or counteracting each other, so any individual factor has limited influence by itself. Therefore, the simulation of the potential vegetation types on the basis of environmental parameters becomes a focus in vegetation ecology. In general, climate factors are mostly considered in the simulation across large landscape (Xu et al., 2012). However, many studies

have taken account of elevation, slope steepness and configuration, aspect, soil, groundwater level and so on at region scale (Dymond and Johnson, 2002 ; Pfeffer et al., 2003). In the view of methodology, relationship is established between environmental variables that are correlated with or surrogate for environmental or resource gradients and species distribution patterns, followed by the application of that model to a geographic database to produce the predictive map, a realization of the model (Franklin, 1995). We selected grassland as an object in the Heihe River Basin. Relationships between grassland distribution and climate factors (precipitation and temperature) or other factors were built in the upper, middle and lower reaches to obtain the potential distributions of grassland based on mapped environmental variables and GIS software. Simulating the potential distribution of grassland in the upper reaches of the Heihe River.

The real grassland distribution is extracted from the land classification from TM images using GIS spatial analysis software. We also obtained the spatial distribution layers of precipitation and temperature using equation (4-1) and equation (4-2). The real grassland distribution layer was used to extract the distribution space of the annual average temperature and precipitation in the upper reach (Figure 4-4), that is, we derived two dimensions' scattergrams about precipitation and temperature during the extent of grassland distribution. The borders of the niche space were delineated by border functions. Figure 4-4 shows the space of temperature and precipitation in the upper reach with the upper border function and the down border function as follows.

$$P_u = -1.464T^2 - 26.12T + 414.4 \tag{4-3}$$

$$P_l = 0.969T^2 - 11.23T + 294.0 \tag{4-4}$$

where P is precipitation; T is temperature. Equation (4-3) is the upper border function, and Equation (4-4) is the down border function (Figure 4-2). Potential spatial distribution of grassland in the upper reach was derived according to the defined niche space by ARC/INFO software.

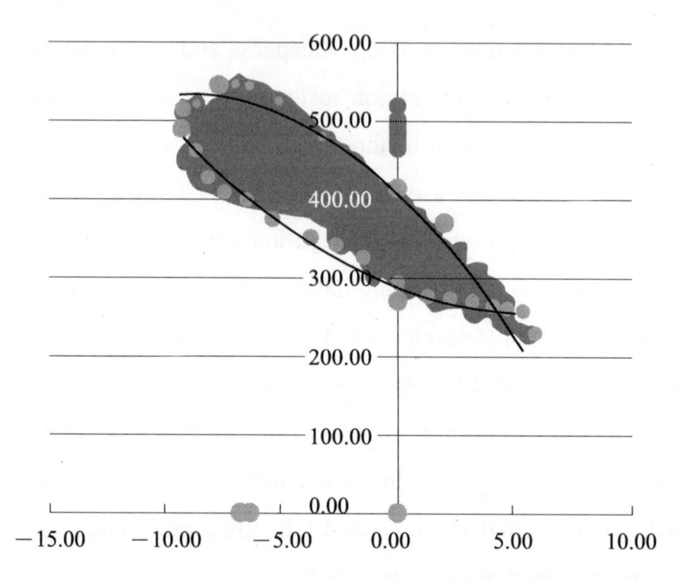

Figure 4-4 Space of annual air temperature and annual precipitation in the upper Heihe River Basin

2. Simulating the potential distribution of grassland in the lower reaches of the Heihe River

Downstream is the extremely arid desert area with annual precipitation less than 50 mm. Although the lower reaches are defined as desert grassland and desert by rainfall threshold (less than 200mm), the area is affected by underground water level. Besides desert grassland, there are riparian grasslands with different coverage. If the underground water level is shallow, grass can grow with high coverage, otherwise, grass grows with lower coverage. We adopted FEFLOW software to simulated underground level. The simulated area was defined as the 10 km buffer where is significantly affected by underground water.

According to field investigation, we developed the relationship between the grass coverage of key species and underground water level as follows.

$$C = \frac{1}{\sqrt{2\pi}\sigma x} \exp\left[-\frac{(\ln x - \mu)^2}{2\sigma^2}\right] \tag{4-5}$$

where C is grass coverage; x is the groundwater depth; μ and σ are the model coefficients. Spatial distribution of groundwater level in the buffer of

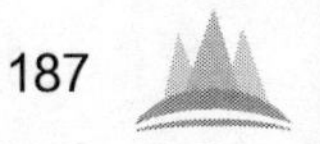

the lower reach was obtained by using FEFLOW model. Then, the potential distribution of grassland was simulated combining with Equation (4-5).

4.2 Results

4.2.1 The spatial distribution of air temperature and precipitation in the upper reaches of the Heihe River

A distribution model of precipitation and temperature is shown in Table 4-1 and Table 4-2. Table 4-1 shows that the accuracy of the annual mean precipitation model is higher to simulate precipitation from May to October, with R^2 values being above 0.8. In the upper reaches of the Heihe River, the precipitation mainly falls in the period from May to September and accounts for 80% of annual total precipitation. So the linear regression equations are considered to be a reasonable choice. Table 4-2 shows that the temperature models have higher accuracy for each month with R^2 values all being above 0.8. In this study, by using regression model and GIS spatial analysis method, the spatial distributions of average annual precipitation and air temperature spatial distributions in the upper reaches of the Heihe River were obtained (Figure 4-5, Figure 4-6).

Table 4-1 Monthly linear regression coefficients and R^2 needed to calculate monthly mean precipitation using altitude (H), latitude (Y) and longitude (X) for the upper reach of the Heihe River

Time	a	b	c	d	R^2
January	−19.811	0.000 260	−0.051	0.231	0.207
February	−70.701	0.001 103	0.221	0.626	0.331
March	−249.545	0.003 390	0.433	2.336	0.406
April	−16.862	0.004 009	−4.289	1.879	0.584
May	408.331	0.009 569	−12.540	0.869	0.810
June	530.716	0.021 000	−13.656	0.016	0.863
July	689.699	0.029 650	−12.485	1.018	0.870
August	495.902	0.018 520	−19.839	2.869	0.879
September	196.940	0.009 100	−15.049	4.003	0.856
October	−5.170	0.002 153	−5.737	2.341	0.841
November	−136.015	0.000 984	0.240	1.283	0.455

Continued

Time	*a*	*b*	*c*	*d*	R^2
December	−81.180	0.000 480	0.493	0.627	0.166
Annual	1 742.001	0.097 260	−87.915	17.197	0.861

Table 4-2 Monthly linear regression coefficient and R^2 needed to calculate monthly mean temperature (SAT) using altitude (*H*), latitude (*Y*) and longitude (*X*) for the upper reach of the Heihe River

Time	*a*	*b*	*c*	*d*	R^2
January	39.168	−0.004 467	−0.935	−0.044	0.893
February	85.077	−0.004 816	−1.262	−0.343	0.961
March	107.647	−0.005 736	−1.319	−0.457	0.981
April	68.609	−0.006 203	−0.248	0.642	0.986
May	19.472	−0.006 538	0.131	0.008	0.983
June	−58.924	−0.006 049	1.230	0.392	0.985
July	−53.779	−0.005 750	1.382	0.294	0.988
August	−28.831	−0.005 816	0.965	0.199	0.985
September	−30.533	−0.005 264	0.596	0.296	0.981
October	−2.415	−0.005 116	−0.101	0.213	0.972
November	64.51	−0.005 668	−1.033	−0.165	0.958
December	41.481	−0.004 511	−0.982	−0.039	0.887
Annual	21.784	−0.00 549	−0.013	0.131	0.981

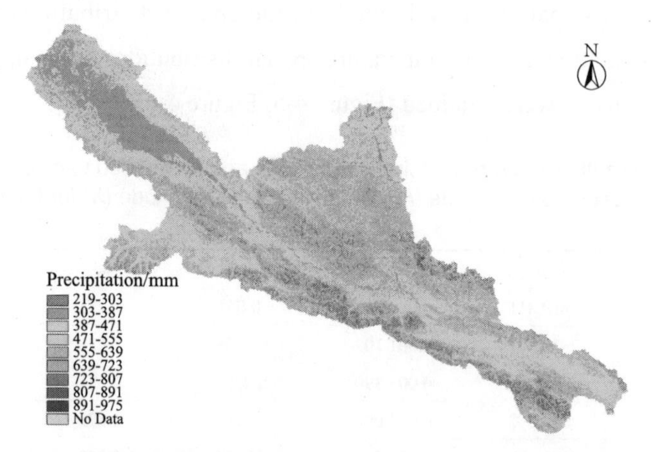

Figure 4-5 The spatial distribution of annual mean precipitation in the upper reach of Heihe River

From Figure 4-5 we can draw some conclusions: ① precipitation decreases significantly from the south to the north. In addition to the obviously latitudinal zonality, longitudinal zonality also exits, precipitation increases with the longitude increase, namely, precipitation increases from west to east. ② rainfall

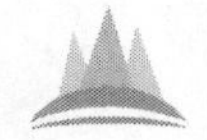

is greatly influenced by topography, and the precipitation increases with the increase of altitude. At the outlet of the Heihe River, precipitation is lower (200-300 mm), but for higher altitude areas, precipitation is higher (400-700 mm). In some areas it can reach more than 700 mm. Figure 4-6 shows: ① the annual average temperature ranges from −13.02 to 6.68℃ with highest value occurring in the mountainous valleys; ② temporally, highest surface air temperature value appears in July (from −3.6 to 21.1℃), and the lowest surface air temperature value can be seen in January (from −23.7 to −9.9℃).

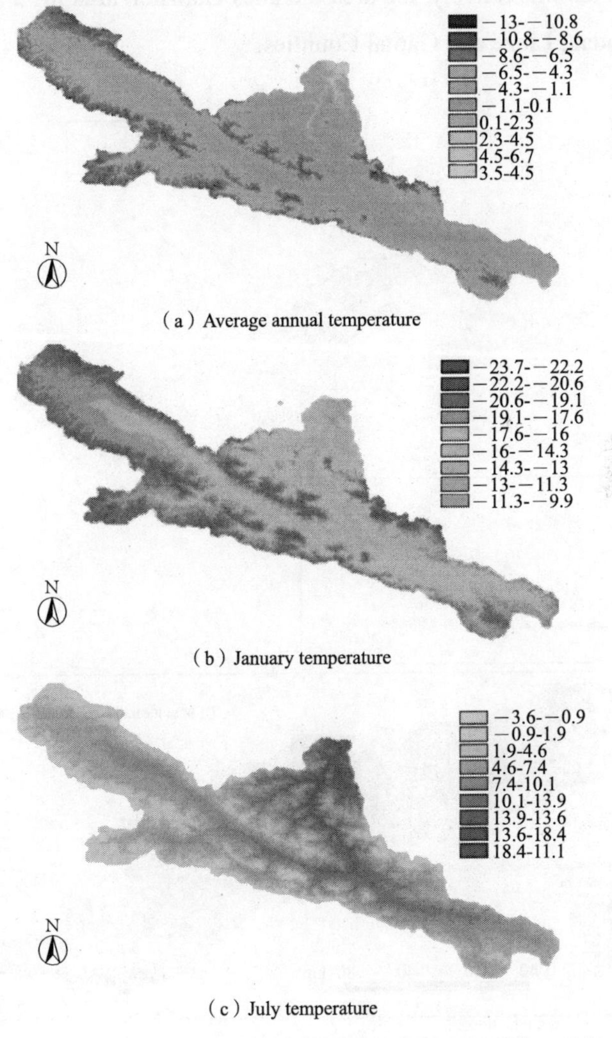

Figure 4-6 The spatial distribution of temperature in different time

4.2.2 The spatial distribution of surface air temperature and precipitation in the middle reaches of the Heihe River Basin

As a main part of Hexi Corridor Plain in the middle of Gansu Province, the middle reaches of the Heihe River (17 000 km^2) is located from 96°42' to 102°00' E and from 37°41' to 42°42' N (Figure 4-7). It is sandwiched between the southern Qilian Mountains and northern Mazong Mountains. It is laid in the junction of low maintains, the oasis and desert. The elevation ranges from 1300 m to 2500 m. Administratively, the area includes Ganzhou area in Zhangye, Sunan, Minle, Shandan, Linze and Gaotai Counties.

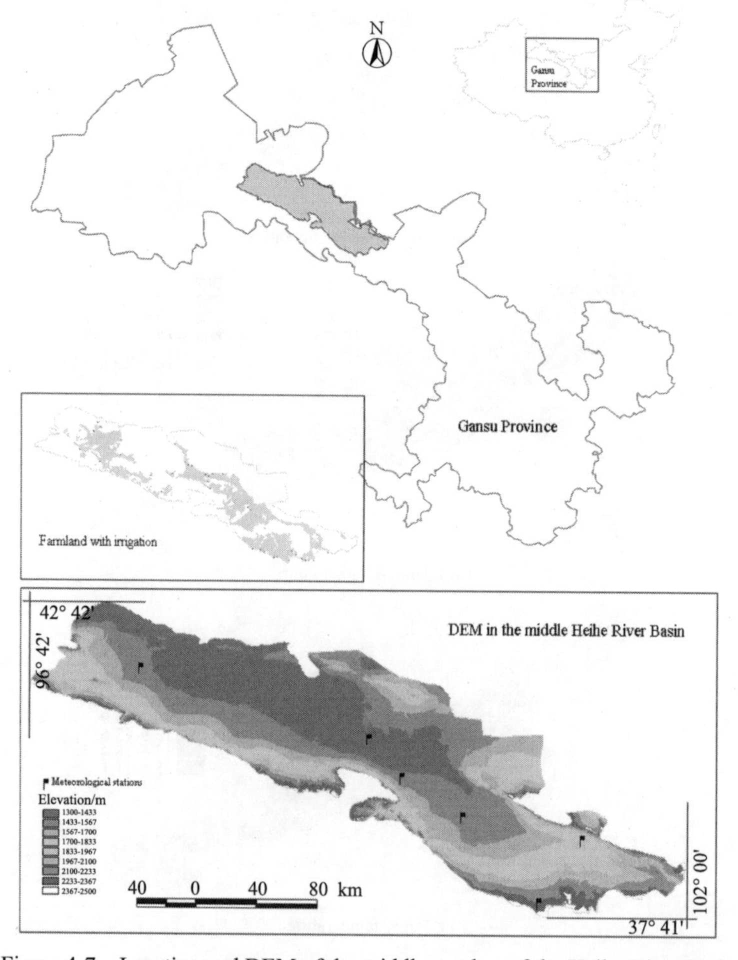

Figure 4-7 Location and DEM of the middle reaches of the Heihe River basin

The spatial distribution of temperature and precipitation is interpolated using OK method (Figure 4-8 and Figure.4-9). From Figure 4-10, temperature increases from southeast to northwest, the range of the average temperature in this area is from 14.5 to 15.8℃. Precipitation is reverse to temperature in the area. The rainfall reduces gradually from the southeast 292.2mm to northwest 86.3mm, which shows a large rainfall gradient (Figure.4-9). The combination of water and heat makes the zonal vegetation in this area be mixture of grassland vegetation and desert vegetation dominated by xerophytic species, or salt tolerance species. According to the field survey, the main species are *nitraria, reaumuria, salsola passerine, ephedra, halogeton arachnoideus* and so on. In addition, there are a lot of artificial vegetation, which depends on irrigation.

Figure 4-8 Distribution of annual air temperature during growing season in the middle reaches of the Heihe River Basin (from 1960 to 2014)

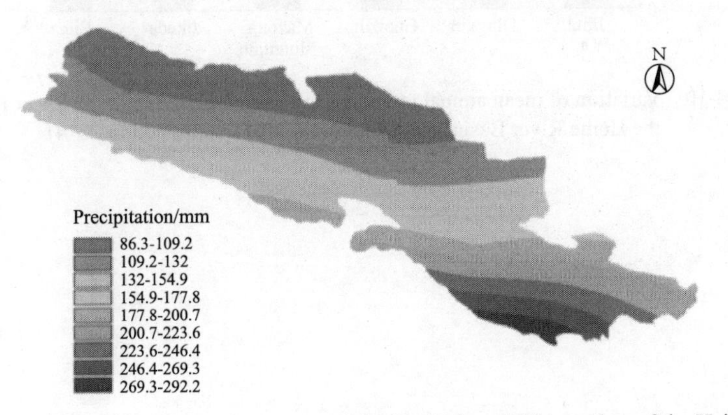

Figure 4-9 Distribution of annual mean precipitation in the middle reaches of the Heihe River Basin (from 1960 to 2014)

4.2.3 The spatial distribution of temperature and precipitation in the lower reaches of the Heihe River

Based on the meteorological station data analysis, the annual average air temperature in the region was no difference except for the higher elevations in the western Mazong Mountains, and annual average temperature ranges from 4.4 to 9.3℃ in the area (Figure 4-10). But there is significant difference in temperature within the year (Figure 4-11). In the area, the annual mean precipitation is less than 100 mm, mainly falling in July, August and September (Figure 4-12). With latitude decreased, precipitation gradually reduces (except Mazong Mountains), and the lowest precipitation appears in northern EJina Banner with less than 50 mm of annual average precipitation (Figure 4-13).

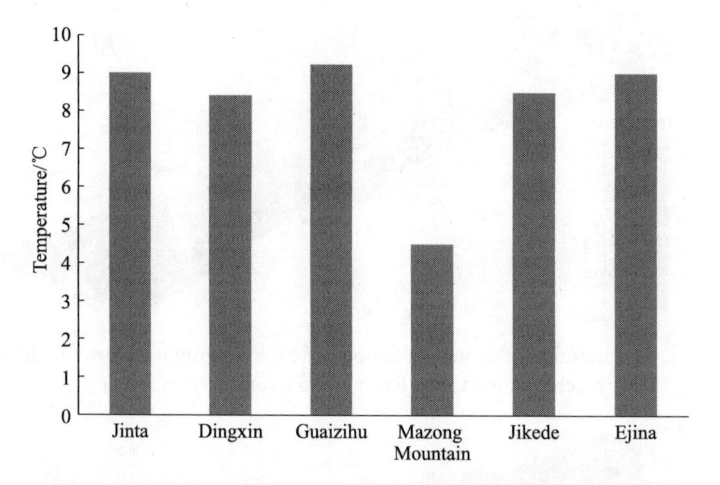

Figure 4-10 Variation of mean annual temperature along the latitude in the lower reaches of the Heihe River Basin (from south to north) (from 1960 to 2014)

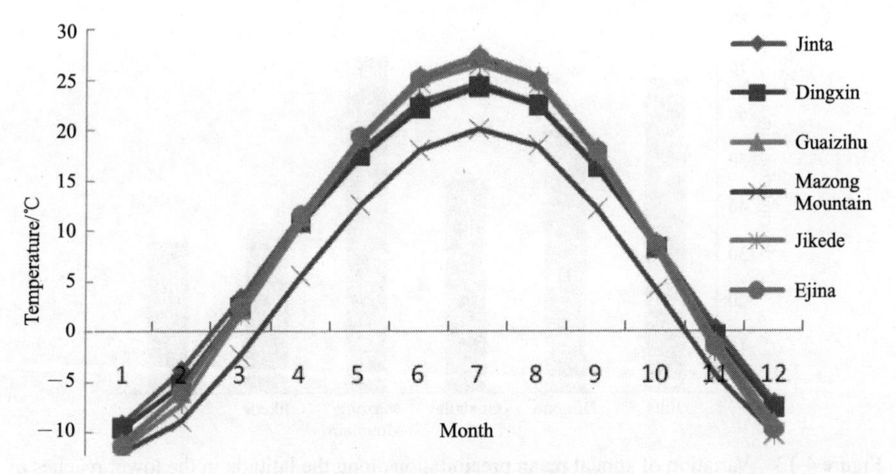

Figure 4-11 Variation of monthly mean temperature in the lower reaches of the Heihe River Basin (from 1960 to 2014)

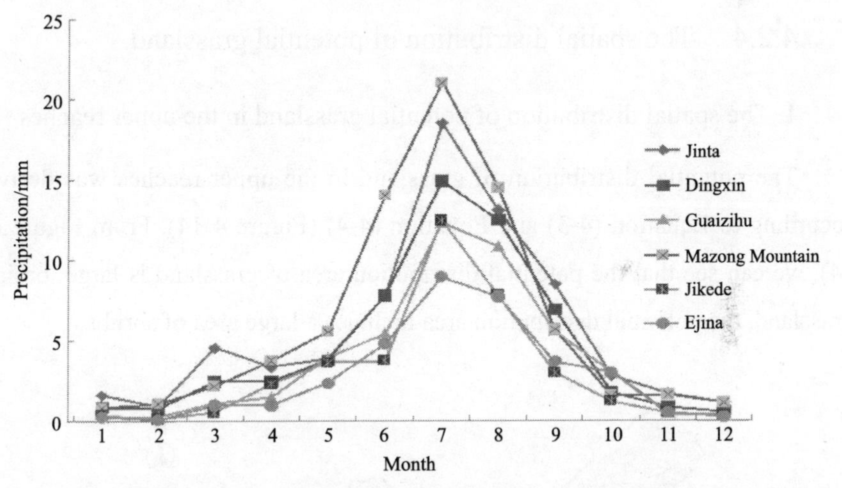

Figure 4-12 Variation of monthly mean precipitation in the lower reaches of Heihe River Basin (from 1960 to 2014)

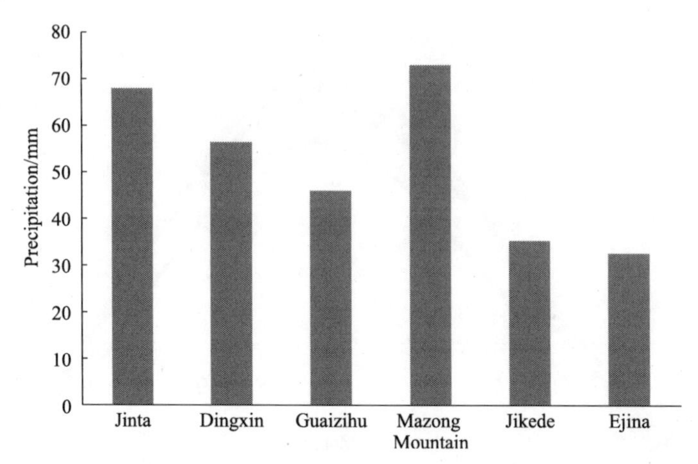

Figure 4-13 Variation of annual mean precipitation along the latitude in the lower reaches of Heihe River Basin (from south to north)

4.2.4 The spatial distribution of potential grassland

1. The spatial distribution of potential grassland in the upper reaches

The potential distribution of grassland in the upper reaches was derived according to Equation (4-3) and Equation (4-4) (Figure 4-14). From Figure (4-14), we can see that the potential distribution area of grassland is large, besides grassland, the potential distribution area includes a large area of shrubs.

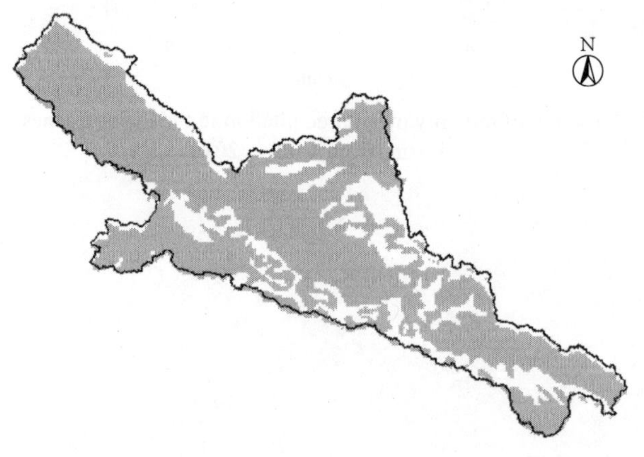

Figure 4-14 Simulation of grassland distribution in the upper reaches of the Heihe River Basin

2. The potential distribution of grassland in the lower reaches of the Heihe River

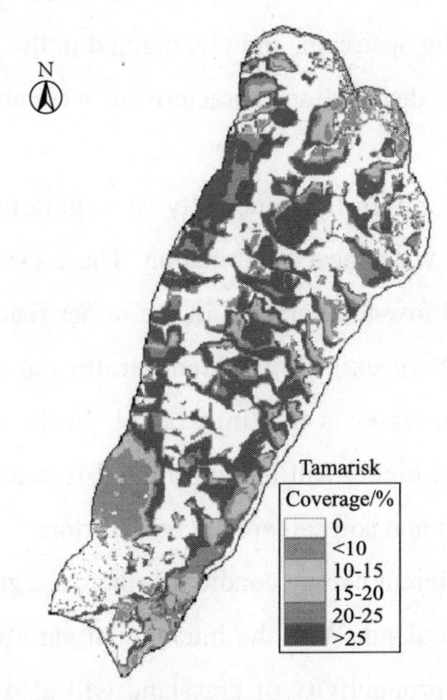

Figure 4-15 Spatial distribution of grassland in the river's buffer in lower reaches of the Heihe River Basin

Using the relationship between groundwater level and vegetation coverage, (i.e., Equation (4-5), and the spatial distribution of the groundwater level, we simulated the potential distribution of grassland in the lower reaches (Figure 4-15). Tamarisk as a main food of livestock in the area, its potential distribution area is smaller than the realistic distribution area, because a large area of tamarisk disappeared mainly due to the decline of groundwater level.

4.3 Conclusions

From the study, we can draw some conclusions as follows.

(1) Because the temperature and precipitation are the key habitat factors of the Heihe River Basin, the spatialization of temperature and precipitation is the precondition of the potential simulation of the grassland vegetation. The

spatialization is restricted by the observation stations deeply. According to terrain condition and derived data, we adopted different methods to analyze the distribution of temperature and precipitation in the Heihe River Basin. Statistical method was used for the upstream, and OK method in the middle reaches. In the downstream, the spatial distribution characteristics were analyzed by station data without spatialization.

(2) The distribution and productivity of vegetation was controlled by different conditions of water-heat combination. There exists the combination of high-precipitation and low-temperature in the upper reaches, and temperature is the limit factor. There exist higher temperature and lower precipitation in the middle reaches, water is the limit factor. In the downstream area, the higher temperature and higher salt amount in soil is combined with the lowest precipitation, and water and soil salt are the limit factors.

(3) Due to the different habitat conditions, there are great differences among the productivity in grassland. With the interannual variation of air temperature and precipitation, the productivity of grassland will also show the interannual changes. Vegetation in the Heihe River Basin has a great dependence and sensitivity to climate change. In the background of future climate change, the type and production capacity of grassland will be changed correspondingly. In grassland management, how to develop mitigation and adaptation strategies for affecting of climate change on grassland is great challenge for managers in future.

(4) The development of the grassland potential model in the Heihe River Basin can not only realize the visualization of potential distribution, but also can be used to analyze the relationship between potential and actual distribution of grassland. Potential distribution can be used as a standard to analyze the change of actual distribution of the grassland at different periods.

(5) The variable of biological geographical model in the potential distribution of grassland is the same as output of the climate model. Coupling the biological geographical model and climate model, we can derive the natural

driving power for grassland change, then, we can analyze the dynamic change of the grassland under climate change scenario.

(6) Potential distribution can be considered as the basis for grassland management. It can be used to distinguish the area where unreasonable expansion and reduction happen. For example, the area of grassland is converted by forest land in the upper reach. The desert and Gobi desert is turned into grassland by using underground and surface water in the middle and lower reaches. Expansion will lead to the enlargement of grassland area which is unstable. The extended grassland will disappear and the original vegetation will recur in case that the humanistic factors disappear. However, grassland reduction due to unreasonable utilization will result in desertification in the lower reach and soil erosion in the upper reach. Even worse, disappearance of the grassland will be irreversible for long time.

Chapter 5

An Analysis of Grassland Variation over Recent 30 Years in the Heihe River Basin

Grassland, the largest terrestrial ecosystem in the world, plays an important role in maintaining ecological balance and guaranteeing sustainable development of social-economic systems. It not only provides materials such as food and medicine, but also supports and maintains the whole system. Ecological services contribute to the whole economic system, both directly and indirectly, and consequently represent part of the total value of our economy system (Bai et al., 2014; Costanza et al., 1997). They are characterized by service process irreversibility, service function irreplaceability, service value externality, service behavior non-marketing, service capital sociality, service space continuity and diversity (Wang et al., 2009). Although these characteristics make grassland a scarce resource supporting people living, they promote inappropriate cultivation and abuse from collection of fuel and medical plants to overgrazing, resulting in severe grassland degradation. Grassland degradation has led to reduction in the capacity of land to perform ecosystem functions and services that support society and development (Li, 1997; Li et al., 2009).

The grassland ecological system, which plays an important role in the Heihe River Basin, is the basis for the development of animal husbandry in the region. The system is influenced by both natural and human activities. The types and coverage of grasslands will change because of the changes in natural conditions

and human disturbance. Analyzing the variation of grasslands and their driving force will benefit the management and maintenance of grasslands, which helps maintaining grassland sustainable development. At present, researchers usually use remote sensing method to study grassland variations in large scale (Jiang et al., 2007; Su et al., 2015; Yu et al., 2006). Based on Landsat TM image data of the multi period U. S. land resources satellites, we have gained the spatial distribution and types conversion of grassland in different periods, which is the primary condition to study the driving force of grassland variation, and also provides basic data for building the relationship between the potential distribution and the actual distribution of grassland (Trabucchi et al., 2014; Yuan, 2015).

5.1 Materials and Methods

5.1.1 Data collection

American Landsat-5 TM and Landsat-7ETM+ image data of 1985, 2000 and 2013 were downloaded. We have collected the DEM of the study area in 90 m resolution. Maps of grassland, land use type, and soil map in the Heihe River Basin were also collected from three provinces.

5.1.2 Methods

We used ERDAS IMAGINE software to process remote sensing images. Grassland classification maps of the Heihe River Basin were interpreted with the method of human computer interactive interpretation in the study period. The classification results were tested by the collected grassland map. According to the classification in different time, we made the statistics of the transition matrix between grassland and other land use types using GIS spatial analyzing technology, and obtained the conversion relationship between the grassland and other vegetation types in different periods. We use the relationship to explain the driving force of the grassland change. At the same time, based on research results from literatures combining with the distribution of grassland types, we calculated the ecological carrying capacity of different grassland types.

5.2 Results

5.2.1 Distribution patterns of grassland in different periods in the Heihe River Basin

Generally, in the lower reaches of the Heihe River Basin, grasslands are dominated by those with low coverage, and the grassland with moderate coverage distributes only on both sides of the river. The grassland with low coverage mainly distributed in the eastern part of Gurinai and the western part of Mazong Mountain (Figure 5-1, Figure 5-2 and Figure 5-3). In the middle reaches of the Heihe River Basin, grassland area is smaller. There is high coverage grassland in Shandan and Minle, and other counties are low coverage grassland. In the up-

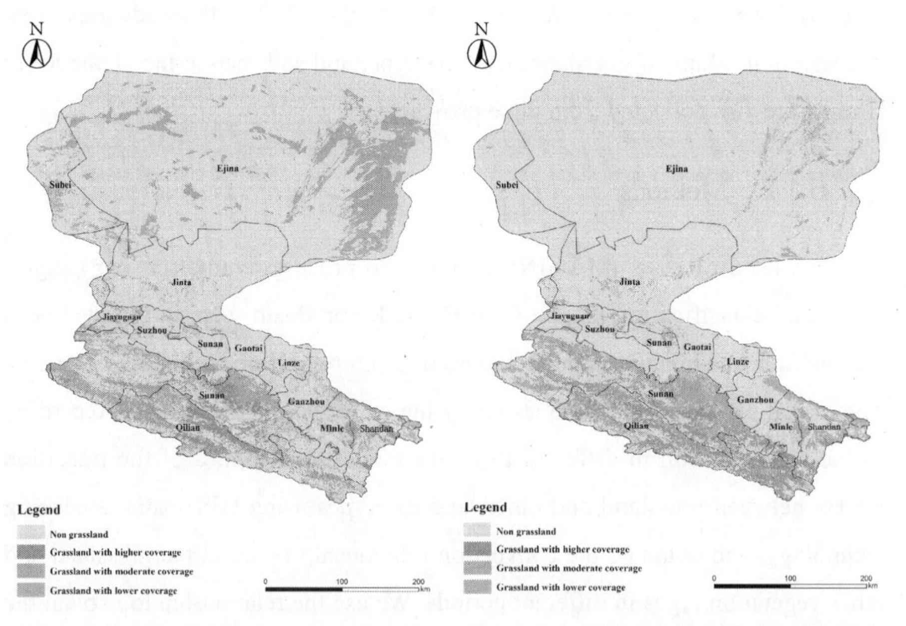

Figure 5-1 The distribution of grassland in the Heihe River Basin in 1985

Figure 5-2 The distribution of grassland in the Heihe River Basin in 2000

per reaches of the Heihe River Basin, high, moderate and low coverage grasslands coexist, and grassland area is larger (Figure 5-1, Figure 5-2 and Figure 5-3). The upper, middle and lower reaches of the Heihe River Basin belong to different administrative regions, so we took the county as a unit to analyze the grassland variation in recent 30 years.

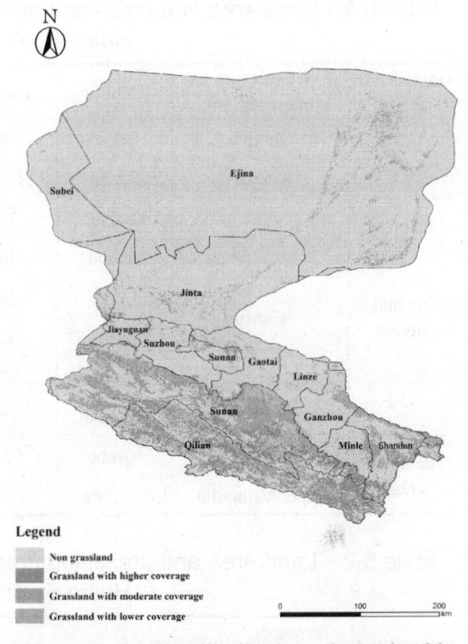

Figure 5-3 The distribution of grassland in the Heihe River Basin in 2013

5.2.2 The spatial variation analysis of grassland in the Heihe River Basin in recent 30 years

We divided the period from 1985 to 2013 into two stages, namely from 1985 to 2000 and from 2000 to 2013. The ratios of grassland to all land area in different period and in different county were displayed in Table 5-1, Table 5-2, and Table 5-3. In the first stage, grassland had an increase trend in the whole basin. Table 5-2 shows that grassland had increased significantly by about 9% in the upper reaches. In the middle reaches, there was an apparent decrease in Suzhou, Gaotai, Linze and Ganzhou whose decrease rates range from 2% to 6.6%, and there was a slight increase in Minle and Shandan. In the lower reaches, grasslands in Ejina Banner had greatly been reduced by 14.21%. In the second stage (Table 5-3), in the upper reaches, grassland in Qilian County had slightly reduced, while grassland in Sunan County increased by 1.67%. In the middle reaches of the basin, grassland in every county had reduced at a different degree except for Ganzhou, Suzhou reduced by 2.73% as the maximum decrement, followed by Minle whose decrease rate is 1.14%. The area of grassland in the downstream increased slightly.

Table 5-1 Land area and grassland area in different administrative divisions in the Heihe River Basin (1985)

Heihe River Basin	Province (autonomous region)	County	Land area/hm^2	Grassland area/hm^2	The ratio of grassland to land area/ %
The up-stream	Qinghai	Qilian	1 100 660.46	604 516.06	54.92
	Gansu	Sunan	1 715 295.58	818 377.54	47.71
The mid-stream	Gansu	Suzhou	338 281.91	40 375.52	11.94
		Gaotai	366 576.47	58 973.03	16.09
		Linze	246 016.29	32 085.80	13.04
		Ganzhou	319 374.05	65 547.55	20.52
		Minle	224 505.13	49 837.67	22.20
		Shandan	299 686.51	136 984.32	45.71
The down-stream	Gansus	Ginta	1 508 327.37	60 904.90	4.04
	Inner Mongolia	Ejina Banner	6 996 865.16	1 176 742.56	16.82

Table 5-2 Land area and grassland area in different administrative divisions in the Heihe River Basin (2000)

Heihe River Basin	Province (autonomous region)	County	Land area/hm^2	Grassland area/hm^2	The ratio of grassland area to land area/%	Variation of grassland area during(1985-2000)/%
The up-stream	Qinghai	Qilian	1 100 660.46	708 513.48	64.37	9.45
	Gansu	Sunan	1 715 295.58	969 067.39	56.50	8.79
The mid-stream	Gansu	Suzhou	338 281.91	33 089.83	9.78	−2.16
		Gaotai	366 576.47	37 025.96	10.10	−5.99
		Linze	246 016.29	15 849.39	6.44	−6.60
		Ganzhou	319 374.05	46 164.23	14.45	−6.07
		Minle	224 505.13	56 777.35	25.29	3.09
		Shandan	299 686.51	154 668.25	51.61	5.90
The down-stream	Gansu	Jinta	150 8327.37	69 277.73	4.59	0.55
	Inner Mongolia	Ejina Banner	6 996 865.16	182 803.50	2.61	−14.21

Table 5-3 Land area and grassland area in different administrative divisions in the Heihe River Basin (2013)

Heihe River Basin	Province (autonomous region)	County	Land area/hm^2	Grassland area/hm^2	The ratio of grassland area to land area/%	Variation of grassland area during 2000-2013/%
The up-stream	Qinghai	Qilian	1 100 660.46	702 835.11	63.86	−0.51
	Gansu	Sunan	1 715 295.58	997 717.83	58.17	1.67

Continued

Heihe River Basin	Province (autonomous region)	County	Land area/hm^2	Grassland area/hm^2	The ratio of grassland area to land area/%	Variation of grassland area during 2000-2013/%
The mid-stream	Gansu	Suzhou	338 281.91	23 849.16	7.05	−2.73
		Gaotai	366 576.47	33 225.42	9.06	−1.04
		Linze	246 016.29	13 583.41	5.52	−0.92
		Ganzhou	319 374.05	46 564.02	14.58	0.13
		Minle	224 505.13	54 216.13	24.15	−1.14
		Shandan	299 686.51	152 392.49	50.85	−0.76
The down-stream	Gansu	Jinta	1 508 327.37	75 125.50	4.98	0.39
	Inner Mongolia	Ejina Banner	6 996 865.16	209 089.37	2.99	0.38

5.2.3 The spatial variation of each county in recent 30 Years

The transformation of grassland to other land use types of each county in two time periods was analyzed as follows.

1. The spatial variation of grassland of Qilian County in recent 30 years

Qilian County, which is located in the upper reaches of the Heihe River Basin, is the main grassland area in the basin. Grassland accounted for 54.92% (in 1985), 64.37% (in 2000) and 63.86% (in 2013) of land, respectively. Grassland had increased by 10% from 1985 to 2000 and kept stable from 2000 to 2013. The transition matrix between grassland and other land use types is shown in Table 5-4. From Table 5-4, we can see that grassland was mainly transformed from shrub, however, grassland is also transformed to swamp and alpine tundra. Conversion of grassland also occurred among grassland types. It is significant that moderate coverage grassland is turned into low coverage grassland. The transition matrix of grassland and other types from 2000 to 2013 can be seen in Table 5-5. Table 5-5 shows that high coverage grassland is turned into marsh. The area of marsh converted to high coverage grassland is larger than that of the reverse direction. And grassland with low coverage turned into the area of alpine tundra is large too. In addition, the shift of different grassland types also happens, the area of high coverage grassland (HCG) converted to the moderate coverage grassland (MCG) and the moderate coverage grassland converted to the low coverage grassland (LCG) is relatively large, which shows that the grassland is degenerating.

Table 5-4 The transition matrix of grassland and other land use types in Qilian County from 1985 to 2000 (unit: km^2)

1985 \ 2000	HCG	MCG	LCG	Swamp	Alpine tundra
Shrub land	685.14	179.44	168.14		
HCG	1022.06	410.40	338.54	242.77	
MCG	537.82	487.16	548.74	186.81	151.02
LCG	309.15	336.07	723.91	109.08	267.59

Note: only listing converted types whose area is more than 100 km^2

Table 5-5 The transition matrix of grassland and other land use types in Qilian County from 1985 to 2000 (unit: km^2)

2000 \ 2013	HCG	MCG	LCG	Swamp	Alpine tundra
Shrub land	2902.56	91.68	27.50		45.06
HCG	33.89	1522.24	78.25		
MCG			2149.28	35.42	
LCG	89.01				

Note: only listing converted types whose area is more than 30 km^2

2. The spatial variation of grassland of Sunan County in recent 30 years

Sunan County is located in the upper reaches of the Heihe River Basin. It has the largest area of grassland in the basin, accounting for 47.71% of the land area in 1985, 56.50% in 2000, and 58.17% in 2013, respectively. During the first period from 1985 to 2000, the grassland increased by nearly 9%, while the area of grassland increased slightly during the second period from 2000 to 2013. The transition matrix between grassland and other land use type in the period of 1985-2000 is shown in Table 5-6. Increased area of grassland is mainly from shrubbery, woodland, salt land, bare rock area, and alpine tundra, which had turned into different type of grassland. At the same time, grassland had been reduced because three kinds of grassland had been converted into canals, flood land, Gobi, and alpine tundra. The shift of three grassland types also happened, the area of

HCG and MCG turned into LCG is larger than that of the reverse direction, and illustrating grassland was degenerating. The transition matrix of grassland and other land use types during 2000-2013 is shown in Table 5-7. Table 5-7 shows that grassland area had increased slightly. Conversion area of Gobi, swamp, alpine tundra and arid plain to low coverage grassland is large, low coverage grassland in high altitude area was also turned into alpine tundra, in the low altitude region, low coverage grassland, also had a certain conversion area into Gobi. The conversion of different grassland types also occurred. By comparing conversion from the transition matrix, we found that grassland had improved in area and coverage.

Table 5-6 The transition matrix of grassland and other land use types in Sunan County from 1985 to 2000 (unit: km^2)

1985 \ 2000	HCG	MCG	LCG	Canals	Flood land	Gobi	Alpine tundra
Woodland	356.14	234.32	339.17				
Shrub land	280.71	160.91	210.33				
Sparse woodland			122.92				
HCG	434.18	331.49	686.41				
MCG	191.02	593.35	1228.44	100.64	65.23	442.61	105.14
LCG	213.64	607.30	1441.79	105.87	89.86	257.06	681.17
Salt land			113.89				
Bare rock land	116.60	334.23	870.96				
Alpine tundra	155.94		306.92				

Note: only listing converted types whose area is more than 50 km^2 of the conversion type

Table 5-7 The transition matrix of grassland and other land use types in Sunan County from 2000 to 2013 (unit: km^2)

2000 \ 2013	HCG	MCG	LCG	Gobi	Alpine tundra
HCG	1606.32	30.75	132.47		
MCG	54.85	2311.45	62.02		
LCG	124.78	82.07	5094.26	31.30	66.27
Gobi			57.18		
Swamp			53.24		
Alpine tundra		32.97	67.97		
Arid land			52.60		

Note: only listing converted types whose area is more than 30 km^2 of the conversion type

3. The spatial variation of grassland of Suzhou District in recent 30 years

Suzhou District belongs to Jiuquan City, in the western part of the middle reaches of the Heihe River Basin, where rainfall is scarce and climate is dry. Grassland is dominated by LCG. Grassland area in 1985, 2000 and 2013 were 40 375.52 hm^2, 233 089.83 hm^2 and 23 849.16 hm^2, respectively, accounting for 11.94%, 9.78% and 7.05% of the regional land area, respectively. During the first period from 1985 to 2000, the grassland had decreased by 18.04%, while the grassland area had decreased by 27.93% during 2000-2013. The transition matrix between grassland and other land use types from 1985 to 2000 is shown in Table 5-8. Table 5-8 shows the conversion of LCG to Gobi, bare soil land and arid land, of which the area to Gobi is the largest. The transition matrix between grassland and other land use types during 2000-2013 is shown in Table 5-9. Table 5-9 shows that LCG mainly is converted to arid land.

Table 5-8 The transition matrix between grassland and other land use types in Suzhou Distinct from 1985 to 2000 (unit: km^2)

1985 \ 2000	MCG	LCG	Gobi	Bare soil	Arid land
LCG	22.60	89.61	147.64	27.35	75.89
Gobi		59.15			
Saline		95.08			

Note: only listing converted types whose area is more than 20 km^2 of the conversion type

Table 5-9 The transition matrix between grassland and other land use types in Suzhou Distinct from 2000 to 2013 (unit: km^2)

2000 \ 2013	MCG	LCG	Gobi	Arid land
MCG	27.20			
LCG		143.52	24.20	62.24
Gobi		31.40		

Note: only listing converted types whose area is more than 20 km^2 of the conversion type

4.The spatial variation of grassland of Gaotai County in recent 30 years

Gaotai County is located in the western part of the middle reaches of the Heihe River Basin, where rainfall is scarce and climate is dry. Grassland is

dominated by low coverage grassland. Grassland area in 1985, 2000 and 2013 were 58 973.03hm^2, 37 025.96 hm^2 and 33 225.42 hm^2, respectively, accounting for 16.09%, 10.10% and 9.06% of the land area of the region. From 1985 to 2000, the grassland had decreased by 37.22%, while the grassland area had decreased by 10.26% during 2000-2013. In the first period of 1985-2000, the transition matrix between grassland and other land use types of land is shown in Table 5-10. As can be seen from Table 5-10: the conversion happens from LCG to sand land, Gobi, marsh land, and arid land, and the conversion area to Gobi is the largest. At the same time, sand land, Gobi and Saline-alkali land turn into LCG, and the shifted area of LCG from Saline-alkali land is the largest. The transition matrix between grassland and other land use types in the second period of 2000-2013 is shown in Table 5-11. From Table 5-11, we can see that the area of grassland had decreased comparing with the first period of 1985-2000, LCG had converted to sand land, saline-alkali land and arid land by 33.00 km^2, 11.15 km^2 and 17.81 km^2, respectively.

Table 5-10 The transition matrix of grassland and other land use types in Gaotai County from 2000 to 2013 (unit: km^2)

1985 \ 2000	MCG	LCG	Flood land	Sand land	Gobi	Marsh land	Arid land
MCG		26.06					
LCG	26.51	114.00	28.57		243.56	41.92	18.73
Sand land		37.70					
Gobi		26.68					
Saline land		71.31					

Note: only listing converted types whose area is more than 10 km^2 of the conversion type

Table 5-11 The transition matrix of grassland and other land use types in Gaotai County from 2000 to 2013 (unit: km^2)

2000 \ 2013	MCG	LCG	Sand land	Saline land	Arid land
MCG	44.28				
LCG		216.83	33.00	11.15	17.81

Note: only listing converted types whose area is more than 10 km^2 of the conversion type

5.The spatial variation of grassland of Linze County in recent 30 years

Linze County, located in the west of the middle reaches of the Heihe River Basin, has a small area with scarce rainfall and dry climate. Grassland is dominated by LCG. Grassland area in 1985, 2000 and 2013 were 32 085.80 hm^2, 15 849.39 hm^2 and 13 583.41 hm^2, respectively, accounting for 13.04%, 6.44% and 5.52% of regional land area. From the first period from 1985 to 2000, the grassland area had reduced by 50.60%, while grassland area had reduced by 14.30% during the second period. During the first period, the transition matrix between grassland and other land use types can be seen in Table 5-12. From Table 5-12, we can know that LCG had been transformed to sand land, Gobi, bare rock land, and arid land, and the area transformed to arid land is the largest, followed by Gobi. At the same time, Saline-alkali land is converted 15.55 km^2 to LCG. During the second period of 2000-2013, the conversion between grassland and other land use is shown in Table 5-13. From Table 5-13, we can see that grassland reduction rate falls comparing with the first period, and LCG is mainly converted to arid land.

Table 5-12 the transition matrix of grassland and other land use types in Linze County from 1985 to 2000 (unit: km^2)

1985 \ 2000	MCG	LCG	Sandy Land	Gobi	Bare rock land	Arid land
MCG		7.23				11.67
LCG	6.27	49.95	19.78	55.78	51.23	90.72
Sandy Land		23.17				
Gobi		21.58				
Saline land	7.34	15.55				

Note: only listing converted types whose area is more than 7 km^2 of the conversion type

Table 5-13 Linze County from 2000 to 2013 grassland and other land use type transition matrix (unit: km^2)

2000 \ 2013	Sparse woodland	HCG	MCG	LCG	Arid land
MCG			10.60		
LCG	15.68			85.22	20.76
Sand land				16.82	
Gobi				13.12	

Note: only listing converted types whose area is more than10 km^2 of the conversion type

6. The spatial variation of grassland of Ganzhou District in recent 30 years

Ganzhou District, located in the middle of the middle reaches of the Heihe River Basin, belongs to Zhangye City. It is a typical agricultural region because of the advantage of irrigation, although rainfall is scarce and climate is dry. Grassland is dominated by LCG. Grassland area in 1985, 2000 and 2013 were 65 547.55 hm^2, 46 164.23 hm^2 and 46 564.02 hm^2, respectively, accounting for 20.52%, 14.45% and 14.58% of regional land area. During the first period of 1985 to 2000, grassland area had decreased by 29.57%, while it had remained stable during the second period of 2000-2013. During the first period, the transition matrix between grassland and other land use types is shown in Table 5-14, which expresses that LCG is mainly converted to Gobi and arid land, and the converted area to Gobi is the largest, followed by arid land and bare rock land. At the same time, Gobi and flood land are converted to LCG in varying degrees. During the second period of 2000 to 2013, the transition matrix of grassland is shown in Table 5-15, which shows that grassland area had remained steady comparing with the first period. LCG is mainly converted to arid land, while sand land and Gobi are converted to LCG in the meantime.

Table 5-14 The transition matrix of grassland and other land use types in Ganzhou District from 1985 to 2000 (unit: km^2)

1985 \ 2000	MCG	LCG	Canals	Flood land	Sand land	Gobi	Bare rock land	Arid land
HCG		10.99						
MCG	8.46	32.70						14.58
LCG	35.53	108.55	11.28	9.30	14.75	190.20	27.17	122.89
Flood land	6.78	15.14						
Sand land		9.02						
Gobi	14.78	46.63						
Bare rock land	47.83	94.23						

Note: only listing converted types whose area is more than 6 km^2 of the conversion type

Table 5-15 The transition matrix of grassland and other land use types in Ganzhou District from 2000 to 2013 (unit: km^2)

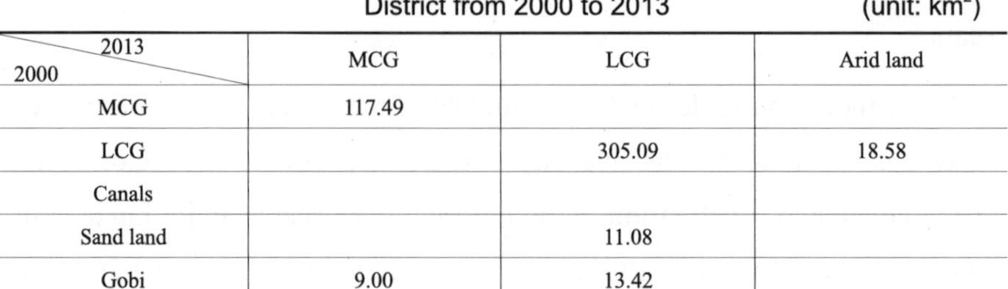

2000 \ 2013	MCG	LCG	Arid land
MCG	117.49		
LCG		305.09	18.58
Canals			
Sand land		11.08	
Gobi	9.00	13.42	

Note: only listing converted types whose area is more than 9 km^2 of the conversion type

7. The spatial variation of grassland of Minle County in recent 30 years

Minle County is located in the eastern part of the middle reaches of the Heihe River Basin, which is a transition region from plain to mountainous area. Because of this, Minle County is an agriculture and pasture interlaced zone in the Heihe River Basin. Rainfall increases with the increase of altitude, which results in various grassland types, including high, moderate and low coverage types. Grassland area in 1985, 2000 and 2013 were 49 837.67 hm^2, 56 777.35 hm^2 and 54 216.13 hm^2, respectively, accounting for 22.20%, 25.29% and 24.15% of the regional land area. The areas of grassland had increased by 13.92% from the first period of 1985-2000, while it had decreased by 4.51% from the second period of 2000-2013. In the first period, the transition matrix between grassland and other land use types is shown in Table 5-16. From Table 5-16, we can see that grassland is mainly transformed from forest land, while grassland is also turned into canals, sand land, Gobi, bare soil land, and arid land. Table 5-17 shows the transfer matrix between grassland and other land use types from second period. From Table 5-17, we can see that the area of grassland remains stable, although grassland is mutually transformed among different grassland types.

Table 5-16 The transition matrix of grassland and other land use types in Minle County from 1985 to 2000 (unit: km^2)

1985 \ 2000	HCG	MCG	LCG	Canals	Sand land	Gobi	Bare soil	Arid hill land	Arid land
Forest land	29.82								
Shrub land	95.52	32.46							

Continued

2000 / 1985	HCG	MCG	LCG	Canals	Sand land	Gobi	Bare soil	Arid hill land	Arid land
Sparse forest land	10.24								
HCG			14.91					10.19	
MCG	13.33	32.87	41.36					44.64	11.42
LCG		19.54	107.16	9.02	11.36	40.38	19.15	30.33	55.52
Flood land			11.77						
Gobi			23.36						
Bare soil land			24.70						
Bare rock land	10.89								
Arid land		8.38	29.08						

Note: only listing converted types whose area is more than 10 km^2 of the conversion type

Table 5-17 The transition matrix of grassland and other land use types in Minle County from 2000 to 2013 (unit: km^2)

2013 / 2000	HCG	MCG	LCG	Arid plain
HCG	144.95		15.46	
MCG		116.19		
LCG			226.67	40.21
Arid land			8.98	

Note: only listing converted types whose area is more than 9 km^2 of the conversion type

8. The spatial variation of grassland of Shandan County in recent 30 years

Shandan County is located in the eastern part of the middle reaches of the Heihe River Basin with larger land area. It is an agriculture and pasture interlaced zone in Heihe River Basin, and the grassland types are various, including high, moderate and low coverage types. Grassland area in 1985, 2000 and 2013 were 136 984.32 hm^2, 154 668.25 hm^2 and 152 392.49 hm^2, respectively, accounting for 45.71%, 51.61% and 50.85% of the regional land area. The grassland area had increased by 12.91% from the first period of 1985-2000 and had decreased by 1.47% from the second period of 2000-2013. During the first period, the transition matrix between grassland and other land use types is shown in Table

5-18. From Table 5-18, we can see that the increase of grassland is mainly due to transformation from forest land, canals, Gobi, bare soil land, arid hill land and arid land in plain, in the meantime, grassland is transformed into Gobi, bare soil land, bare rock land, hill arid land and arid land in plain. During the second period, the transition matrix between grassland and other land use types is shown in Table 5-19, which shows the area of grassland remains stable comparing with the first period, although grassland is transformed mutually among different grassland types.

Table 5-18 The transition matrix of grassland and other land use types in Shandan County from 1985 to 2000 (unit: km²)

2000 / 1985	HCG	MCG	LCG	Flood land	Sand land	Gobi	Bare soil land	Bare rock land	arid land
Forest land	53.22	22.43	32.12						
Sparse forest land	15.60	9.68	15.62						
HCG		36.03	101.78			18.45	13.38		
MCG		128.15	280.96			33.45	9.19	9.33	20.01
LCG		82.89	383.75	15.15	11.90	62.98	68.15		32.17
Flood land		14.66	27.00						
Gobi		35.09	160.06						
Bare soil land			11.84						
Arid hill land			30.40						
Arid land			32.72						

Note: only listing converted types whose area is more than 10 km² of the conversion type

Table 5-19 The transition matrix of grassland and other land use types in Shandan County from 2000 to 2013 (unit: km²)

2013 / 2000	HCG	MCG	LCG	arid hill land	arid land
LCG	84.48				
MCG		341.75	7.07		
LCG			1053.34	7.01	22.37
Arid land			11.13		

Note: only listing converted types whose area is more than 9 km² of the conversion type

9. The spatial variation of grassland of Jinta County in recent 30 years

Jinta County, located in the south of the downstream in the Heihe River Basin, has very dry climate, where grassland area is less than 5% of the regional land area, and is dominated by LCG. Grassland area in 1985, 2000 and 2013 are 60 904.90 hm^2, 69 277.73 hm^2 and 75 125.50 hm^2, respectively, accounting for 4.04%, 4.59% and 4.98% of regional land area. The grassland area had increased by 13.75% from 1985 to 2000 (the first period), and had increased by 8.44% from 2000 to 2013 (the second period). During the first period, the transition matrix between grassland and other land use types is shown in Table 5-20. From Table 5-20, we can see that the increase of LCG is mainly due to the conversion from sparse forest land, shrub land, sand land, canals, Gobi, saline-alkali land, bare soil land and arid land, and low LCG is transformed into sand land, Gobi, bare soil land, bare rock land, and arid land in the meantime. During the second period, the transition matrix between grassland and other land use types is shown in Table 5-21, which shows that the grassland area is mainly transformed from sand land, at the same time grassland is also transformed into sand land and arid land.

Table 5-20 The transition matrix of grassland and other land use types in Jinta County from 1985 to 2000 (unit: km^2)

1985 \ 2000	HCG	MCG	LCG	Sand land	Gobi	Bare rock land	Arid land
Shrub forest land			22.52				
Sparse forest land			10.21				
LCG			175.52	154.71	98.35	80.27	27.82
Flood land			13.29				
Sand land			135.80				
Gobi			86.98				
Saline-alkali land			158.09				
Bare soil land			16.52				
Arid land			23.10				

Note: only listing converted types whose area is more than 10 km^2 of the conversion type

Table 5-21 The transition matrix of grassland and other land use types in Jinta County from 2000 to 2013 (unit: km^2)

2013 / 2000	MCG	LCG	canals	Sand land	Gobi	Saline-alkali land	Wetland	Bare soil land	Arid land
Sparse forest land		8.62							
MCG	14.34								
LCG	19.18	374.27	6.36	64.02	12.20	10.03	47.66	8.32	104.42
Flood land		8.07							
Sand land		230.97							
Gobi		27.92							
Saline-alkali land		6.90							
Arid land		9.36							

Note: only listing converted types whose area is more than 10 km^2 of the conversion type

10. The spatial variation of Ejina Banner grassland in recent 30 years

Ejina Banner is located in the low reaches of the Heihe River Basin with the largest area, where annual mean precipitation is less than 50 mm, so climate is extremely dry. This area is dominated by LCG. In the year of 1985, 2000 and 2013, the grassland area is 1 176 742.56hm^2, 182 803.50hm^2 and 209 089.37hm^2, respectively, accounting for 16.82%, 2.61% and 2.99% of the regional land area. During the first period of 1985-2000, the grassland area had decreased fast, reducing by 84.47%. However, during the second period of 2000-2013, it had increased by 14.38%. During the first period, the transition matrix between grassland and other land use types is shown in Table 5-22. Table 5-22 shows that vast areas of LCG had been converted to sand land, Gobi, saline-alkaline land and bare rock land. Besides, riparian areas, sand land, Gobi, saline-alkaline land and arid land had also been converted to LCG, and meanwhile, LCG had been converted to sand land, Gobi, bare rock land and arid land in various degrees. The transition matrix between grassland and other land use types for the second period has been displayed in Table 5-23, which shows that increased grassland area is mainly from sand land.

Table 5-22 The transition matrix of grassland and other land use types from 1985 to 2000 in Ejina Banner (unit: km²)

2000 / 1985	Forest land	Spare forest land	HCG	MCG	LCG	Riparian areas	Sands land	Gobi	Saline-alkaline soil land	Bare rock land
Shrubbery			25.85	102.86	134.54					
MCG		25.27		75.12	97.63		74.98	47.74	16.09	
LCG	23.06	47.11		169.42	609.21	171.62	3822.39	5037.42	561.89	811.08
Sand land				18.97	100.86					
Gobi				12.07	159.57					
Saline-alkaline soil land				37.17	168.19					

Note: only listing converted types whose area is more than 20 km² of the conversion type

Table 5-23 The transition matrix of grassland and other land use types in Ejina Banner from 2000 to 2013 (unit: km²)

2013 / 2000	Open forest land	HCG	MCG	LCG	Sand land	Gobi	Saline-alkaline soil land	Bare rock land
HCG		36.32						
MCG	10.36		354.74	27.56	13.46			10.68
LCG	21.55		37.27	1033.18	105.11	50.49	28.73	18.67
Sand land		32.58	97.91	247.60				
Gobi			14.58	63.46				
Saline-alkaline soil land			9.19	50.19				

Note: only listing converted types whose area is more than 10 km² of the conversion type

5.2.4 The analysis of grassland carrying capacity

Oldeman et al. (1981) pointed out that 34.5% land degradation was caused by overgrazing in the world, more serious in the arid and semi-arid areas which account for 80% of the world grassland area. How to remain proper grazing intensity in order to protect from grassland degradation? How to define proper livestock carrying capacity? These questions have drawn much attention from decision makers (Zhang et al., 2005). Under the background, the grassland carrying capacity was put forward.

Grassland carrying capacity is defined as the amount of livestock carried by a certain area of grassland without damaging grassland resources in a specific

period, which can be divided into production carrying capacity and nutrition carrying capacity. In terms of production carrying capacity, 50% of grass utility ratio is used to calculate production carrying capacity (Wang and MA, 1994), because the other 50% is remained for protecting soil from erosion and ensuring restoration of grass vegetation in drought as well as meeting the needs of regeneration, insects, soil invertebrates and wild animals. Grassland carrying capacity can be calculated as follows (Holechek et al., 2001).

$$C_i = \frac{F_i(1-E_i)(1-\mathrm{DW})(1-\mathrm{DS})}{G} \tag{5-1}$$

$$S = \sum_{i=1}^{n} C_i A_i \tag{5-2}$$

where C_i is grassland carrying capacity per unit area at grassland type i (the amount of livestock carried per unit area); F_i is per unit yield at grassland type i; E_i is the allocation of forage root biomass in grassland type i; DW is the reduced proportion of carrying capacity with the increase of distance from water source; DS is the reduced proportion of carrying capacity with the increase of slope; G is the amount of grass biomass met by a sheep; S is the amount of livestock carried per unit area at a certain area; A_i is area of grassland type i; and n is number of grassland types.

In the Heihe River Basin, the climate has differences that result in many kinds of grassland with different forage biomass. In terms of grassland coverage, the grassland can be divided into HCG, MCG and LCG. According to the present status of grassland in the Heihe River Basin, it's considered that the amounts of forage biomass used to meet the needs of insects, soil invertebrates and wild animals are 15%, 25% and 35% of the total forage biomass in HCG, MCG and LCG, respectively. The adjusting values of DW and DS are listed in Table 5-24. In mountainous area of the Heihe River Basin, sheep are Tibet sheep with larger body, whose daily amount of consumption is defined as 5 kg. Mongolia sheep lived in plain desert have smaller body, whose daily amount of consumption is defined as 3 kg. Sheep lived in the farming pastoral region and agricultural region, whose daily consumption amount is defined as 4 kg. According to the

results of field investigation, the grass yield of grassland in different regions is shown in Table 5-25.

Table 5-24 Adjustment values of carrying capacity regarding to distance to water source and slopes

Distance to water Source/km	Reduced grazing capacity/%	Slopes/%	Reduced grazing capacity/%
0-1.6	0	0-5	0
1.6-3.2	50	5-15	30
>3.2	100	15-25	50
		25-35	70
		>35	100

Data Source: Zhang et al.(2005)

Table 5-25 Natural grassland classification, distribution and productivity in the Heihe River Basin

Grassland types	Distribution area	Coverage/%	Productivity/(kg/hm²)
Low humidity Swamp grassland	Riparian area in the lower and middle reaches	High coverage (>75%)	6000-7500
High-cold Swamp grassland	High mountainous area in the upper reaches	High coverage (>75%)	375-450
Low humidity meadow grassland	River and lake beach in the lower and middle reaches	Moderate coverage (40%-80%)	6000-5500
Plain desert grassland	Desert area in the lower and middle reaches	Low coverage (<10%)	<450
Plain semi-desert Grassland	Pediment plain	Moderate and low coverage (5%-30%)	750-1200
Desert riparian forest shrub grassland	Open forest land and shrubbery in the lower reaches	High coverage (>50%)	750-5700
Mountainous desert grassland	Low mountains and hills in the lower and middle reaches	Low coverage (5%-20%)	150-1050
Mountainous semi-desert grassland	high mountain valleys in the upper reaches	Moderate coverage (30%)	750-1200
Mountainous grassland	Elevation range from 2100 m to 3800 m	High and moderate coverage (30%-70%)	975-1425
Mountain meadow grassland	Elevation range from 2800 m to 3800 m	High coverage (>75%)	1950-2250
Piedmont shrub meadow grassland	Elevation range from 3000 m to 3700 m	High coverage (>50%)	1350-2250

According to Equation (5-1) and Equation (5-2), we can calculate carrying capacity of the Heihe River Basin in different administrative regions in different periods (Figure 5-4, Figure 5-5, Figure 5-6) combining with results of previous researches, the data of field investigation and grassland area.

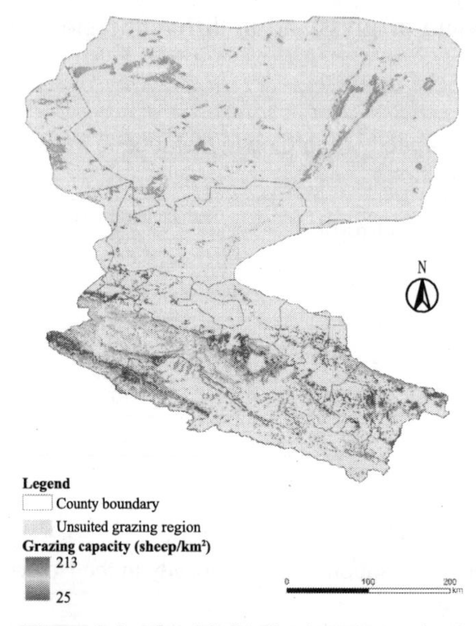

Figure 5-4 Spatial distribution of grassland carrying capacity in the Heihe River Basin (1985)

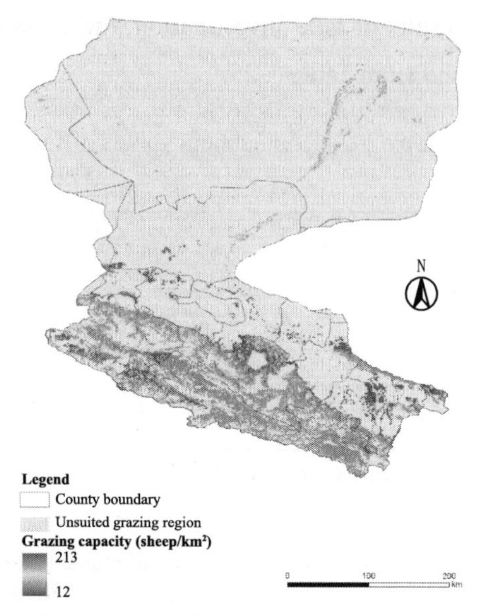

Figure 5-5 Spatial distribution of grassland carrying capacity in the Heihe River Basin (2000)

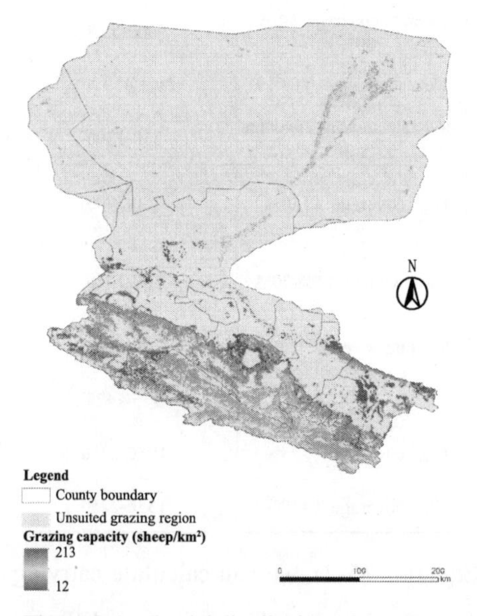

Figure 5-6 Spatial distribution of grassland carrying capacity in the Heihe River Basin (2013)

5.3 Conclusions

We used remote sensing data for grassland classification in the Heihe River Basin during different periods, the results show that grassland types had greatly changed in the last 30 years. In the first period of 1985-2000, the grassland types had increased in the upper reaches and had decreased in the middle and lower reaches, respectively. In the second period of 2000-2013, the grassland area of the whole basin kept stable.

From the transition matrix be-

tween grassland and other land use types, we can obtain that the main driving force of grassland variation is human factor, that is, the policy adjustments result in reciprocally transformation between grassland and other land use types obviously.

Under different natural conditions, the productivity of grassland is different, and in turn, the carrying capacity of grassland is also different. In the management of grassland, the development of animal husbandry should strictly be in accord ance with the principle of the balance between grass and livestock. Overgrazing will lead to the deterioration of grassland and the unsustainable utilization of grassland resource.

Distribution patterns of grassland in the Heihe River Basin based on remote sensing data interpretation lay the foundation for the potential grassland simulation and are beneficial to building the relationship between potential grassland and real grassland.

Chapter 6

An Analysis of Grassland Management Policies Based on An Opinion Survey of Farmers and Herdsmen

6.1 Questionnaire Design

Named as *Research into Grassland Management Policies in the Heihe River Basin: A survey of Living and Production Mode of Farmers and Herdsmen and the Status Quo of Policy Implementation in the Region*, the survey is aimed at having a better understanding of living and production mode of farmers and herdsmen as well as the implementation of related policies on grassland management. Its respondents are farmers and herdsmen in the Heihe River Basin, who are at the most grass-root level of grassland management and protection. The theme of the survey is an evaluation of the grassland management policies in the Heihe River Basin by farmers and herdsmen. A questionnaire was designed to investigate the living standards and lifestyle of farmers and herdsmen so as to obtain the first-hand information in this regard. Results and policy recommendations from the survey would be of basic reference to further policy-making.

6.1.1 Questionnaire content

The questionnaire of the survey was drafted and designed based upon in-depth discussions among research members as well as suggestions from experts in the fields. It consists of four sections, including basic information of respondents, economic status of respondents, current situation of grasslands in the basin, and grassland management policies in the basin. There were several questions in each section, which were summarized in Table 6-1.

Table 6-1 Summary of main content of the questionnaire

Section	Content
Basic information of respondents	Ethnic groups, gender, age, educational level
Economic status of respondents	Household characteristics: household size and labor force, and quantity of durable goods Household income: current status of animal husbandry and the related income, current status of crop cultivation as well as the related income, major income sources, net income per annum, income fluctuation Household expenditure: expenditure on animal husbandry and crop cultivation, living consumption expenditure
Current situation of respondents	Causes on fluctuation of temperatures, precipitations and grassland ecological status
Discussion on current implemented policies of grassland	Evaluation of grassland ecological protection policies Rewards and compensation on grassland ecological protection Results of current implemented protection policies and measures Punishments for behaviors causing grassland deteriorating Major reasons behind the violation of management rules Impacts of grassland protection policies or measures on incomes Impacts of grassland protection policies or measures on ecological environment Feasible measures for farmers and herdsmen to raise their household incomes Key issues of future government management

In the first section, basic information of respondents would indicate the variances among ethnic groups and cultures. It would be a main factor influencing the respondents' recognition of grassland management policies. Therefore, during the survey, the questionnaire was handed out according to the weightings of ethnic groups. Older respondents with long work experience of animal husbandry in the pastoral areas would be selected for the survey. Also, variances and divisions of genders and cultures were taken into consideration in order to ensure the authenticity and validity of the survey results.

In the second section, economic status of farmers and herdsmen is regarded as the key factor influencing their attitudes to the policies. Responses to, and statistical analysis on, questions concerning major income sources, income amount, fluctuation of income and status of expenditure would be the underlying foundation for further policy making, implementation and causal analysis on policy effects.

The third section was designed to figure out the natural causes for the environmental alterations of grasslands in the Heihe River Basin due to the implementation of related management policies.

The fourth section concerns policy implementation, policy effects as well as their causes. And the key issues of the current policies were set forth for the final analysis. An analysis in the fourth section would provide the statistical and practical support for the improvement on the further policy-making.

6.1.2 Survey plan

1. Mode of survey

In order to ensure its authenticity and validity, the research was mainly conducted and arranged as a field survey. Researchers and team members from the Institute of Policy and Management, CAS and Lanzhou University carried out a field trip to the related counties and areas. By face-to-face interviews, farmers and herdsmen were willing to raise their voices and share their perceptions based upon their daily farming and husbandry work. First-hand data and information were collected by this mode of survey.

2. Target areas for questionnaire

This field survey covered all the counties in the Heihe River Basin, which resulted in a large number of samples. A total of 1200 questionnaire copies were handed out. Among the 1150 responding questionnaires, 1017 were valid (754 valid respondents from Gansu Province, 148 valid respondents from Inner Mongolia Autonomous Region, and 115 valid respondents from Qinghai Province). Distribution of the questionnaire among counties and related areas

was listed in Table 6-2. Geographical distribution of the questionnaire was spatially equal, and the daily work of respondents was mainly focused upon crop cultivation and animal husbandry. It is beneficial and persuasive to discover the current status of grassland management in the Heihe River Basin.

Table 6-2 Distribution of the questionnaire among counties and related areas

Provinces(autonomous region)	Areas	Counties	Copies of questionnaire
Qinghai	Northern Tibetan Autonomous Prefecture	Qilian County	115
Gansu	Zhangye City	Sunan County	138
		Shandan County	87
		Minle County	109
		Ganzhou District	18
		Linze County	141
		Gaotai County	112
	Jiuquan City	Jinta County	149
Inner Mongolia	Alashana League	Ejina Banner	148

6.2 Survey Results

6.2.1 Characteristics of respondents

The survey covered farmers and herdsmen of all the counties in the Heihe River Basin. As to ethnic characteristics, the majority of respondents were Han nationality while a substantial proportion of them were ethnic minorities such as Tibetans, Mongolians and Yugurs. In terms of the respondents' gender characteristics, the majority of them were male with the male-to-female ratio up to 3∶1. Regarding their age distribution, most of the respondents were in the age group of 21-60. More than 60% of them were in the age group from 31 to 50. As to their educational level, the majority of them had been to primary school or high school (or technical secondary school). Accordingly, distribution of respondents was pretty rational.

6.2.2 Household characteristics of respondents

1. Basic characteristics of households

The number of working-age people and durables in a household are described as the basic characteristics of households (Figure 6-1). As to household size, the survey shows that the average number of people in a household was four. The average number of working-age people in a household was 2.5. The average number of family members doing crop farming, animal grazing and migrant work in a household was 2, 1.8 and 1, respectively. It shows that more people doing farming than doing animal grazing, and on average, every family had one member working away from home.

When it comes to durables in a household (Figure 6-2), they were classified according to different functions: transporting (including electric vehicles, motorcycles, and automobiles), production (including tricycles, four-wheel vehicles, tractors, and water pumps) and living purposes (including refrigerators, TV sets, and washing machines). A statistical analysis shows that the proportion of transporting and production durable goods in a family was pretty low while that for living purposes was rather high. It reveals that the living standards of farmers and herdsmen have improved substantially; however, their production and transportation facilities are still quite backwards and need improvement.

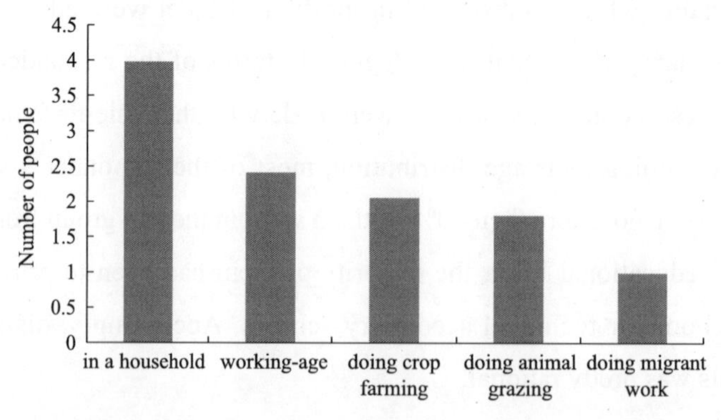

Figure 6-1 Household size and its composition in the Heihe River Basin (units: number of people)

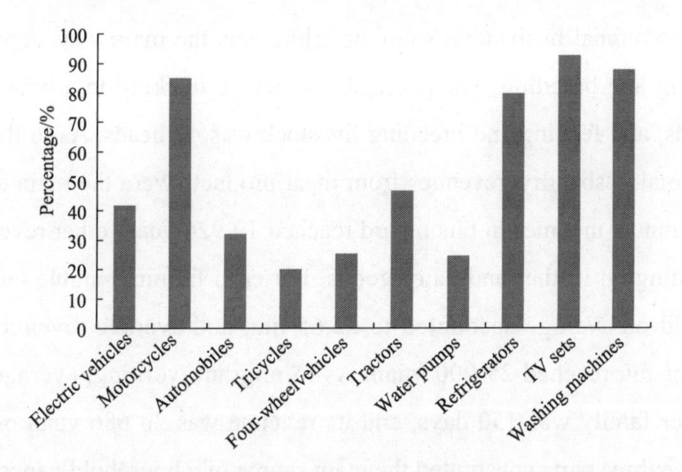

Figure 6-2 Proportions of durable goods in a household in the Heihe River Basin

2. Household incomes

According to the respondents' feedback, income sources of a household in the region were comprised of three parts: animal grazing, farming and migrant working (Figure 6-3). On average, their incomes from animal grazing, farming and migrant working were 33 310 yuan, 24 968 yuan, and 20 686 yuan, respectively. Therefore, animal husbandry was the major household income source while income of migrant working became an important income sources. For the net annual income per household, three quarters of them were between 10 000 and 70 000 yuan; more than 30% were between 30 000 and 50 000 yuan. That is to say, most of the households were low and middle-income families.

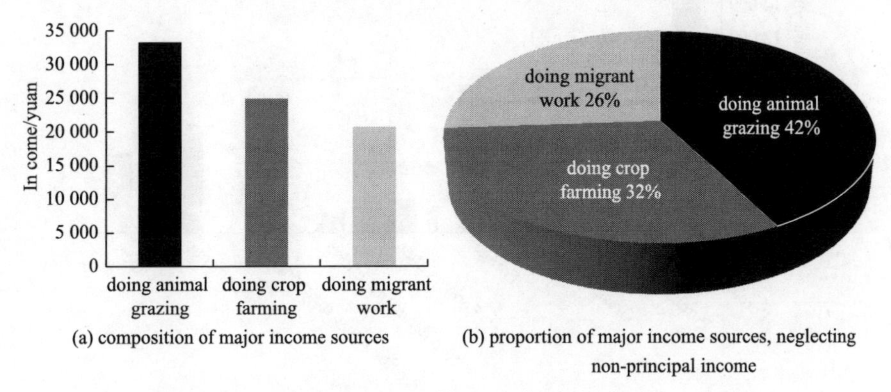

(a) composition of major income sources

(b) proportion of major income sources, neglecting non-principal income

Figure 6-3 Income sources of households in Heihe River Basin

As to animal husbandry, stock breeding was the main way, supplemented by feeding and breeding. The average number of stocking in a household was 143 heads, and feeding and breeding livestock was 42 heads. As to the incomes from animal husbandry, revenues from meat products were the main source and average annual incomes in this regard reached 19 926 yuan, other revenues were from trading on leather and dairy goods. For crop farming, arable land of each household on average accounted to 22.65 mu, and average revenue from the yields per mu reached 25 000 yuan. As to migrant working, average working period per family was 150 days, and its revenue was 20 686 yuan on average. The above three parts constituted the main source of a household's income.

3. Household expenditure

Household expenditure fell into three major items: animal husbandry, crop-farming, and living consumption. Generally, the total expenditure of a household was 95 761 yuan on average, including 48 122 yuan on animal husbandry, 22 431 yuan on farming and 25 206 yuan for living consumption. Regarding the expenditure on animal husbandry (Figure 6-4), those on pups and forage took the lion's share: 16 708 yuan and 9025 yuan respectively. As to expenditure on farming, the payment on hiring laborers (Figure 6-5) took the lion's share, 5653 yuan which was followed by the expenditure on chemical fertilizer, 4287 yuan.

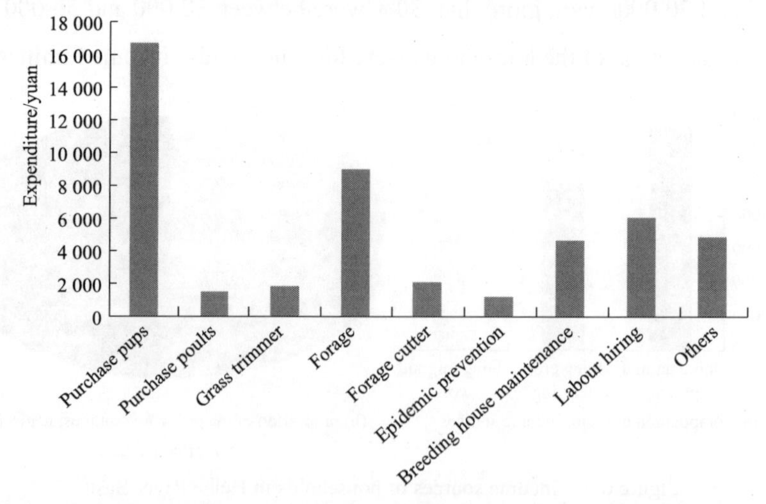

Figure 6-4 Household expenditure on animal husbandry in the Heihe River Basin

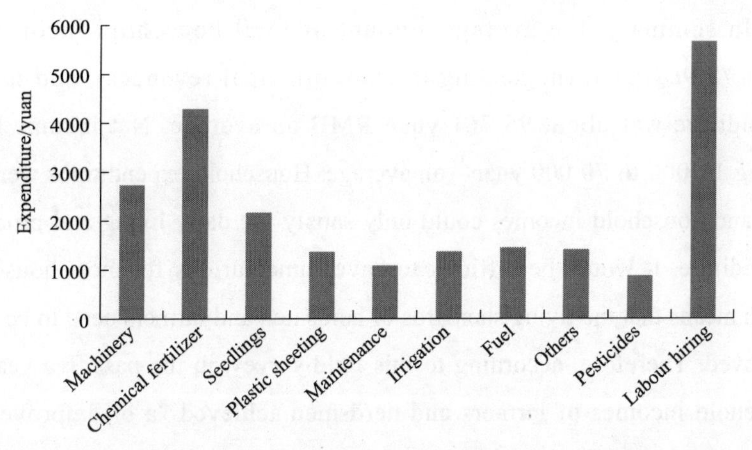

Figure 6-5 Household expenditure on crop farming in the Heihe River Basin

The fact that the expenses of doing animal husbandry top a household's expenditure means its productive cost was pretty high, especially for the purchases of pups, while the labor investment was fairly low. Additionally, the cost of farming was pretty low while the cost of labor-force was rather high. The total expenditure on living consumption was pretty high, even higher than that of farming. It means that the consumption level of the locals had been raised along with the improvement of living standards. However, the proportion of food consumption was also pretty high and so was the Engle Coefficient, which means the improvement of living standards was rather limited. Expenditure on educational purposes was pretty high, revealing that farmers and herdsmen paid much attention on the children's education.

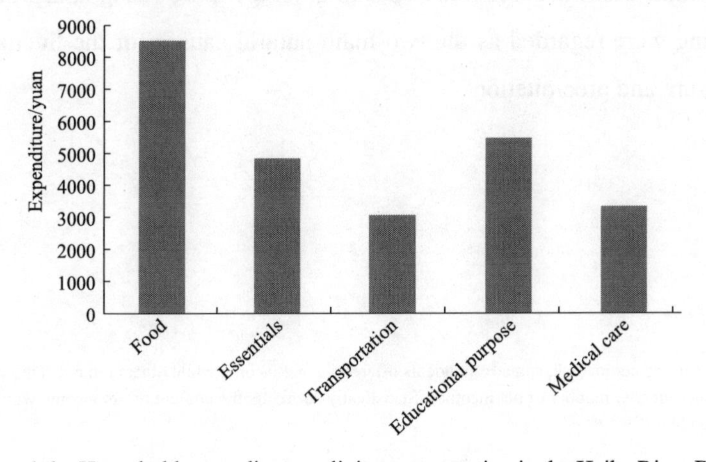

Figure 6-6 Household expenditure on living consumption in the Heihe River Basin

In summary, the average amount of total household incomes was about 78 964 yuan (neglecting the non principal revenues), and the total expenditure was about 95 761 yuan RMB on average. Net income ranked among 10 000 to 70 000 yuan[①] on average. Household expenditures were quite high and household incomes could only satisfy the daily living and production expenditure. It would be difficult to have some surplus for most households, which means that the living standards of herdsmen and farmers need to be further improved. Therefore, according to this field survey, in the past five years, the household incomes of farmers and herdsmen achieved "a bit improvement". Improvement on living standards, production level and household income seems unobvious. Most of the farmers and herdsmen were not satisfied with their economic status.

6.2.3 Survey on grassland management policies

Temperature and precipitation are the two main natural factors influencing the ecological environment of the Heihe River Basin. Some questions were designed to investigate farmers' and herdsmen's recognition on the fluctuation resulted from the two natural factors over the recent 10 years (Table 6-3). Regarding temperature, nearly 60% of the respondents thought that there was "a slight ascent"; as to precipitation, nearly 35% of the respondents thought there was "a slight descent". Generally speaking, temperature rising and precipitation decreasing were regarded as the two main natural causes for the fluctuations of temperature and precipitation.

① As to the net income, some respondents offered the net income while others did not. This would finally lead to a much smaller number of net incomes. Statistically speaking, the amount of net income was smaller than that of expenditure.

Table 6-3 Natural causes for the environmental alterations in the Heihe River Basin

Natural causes for the environmental alterations	Options	Frequency	Relative frequency
Temperature fluctuation	a greater ascent	140	13.8
	a slight ascent	603	59.3
	no obvious change	122	12.0
	a slight descent	84	8.3
	a greater descent	13	1.3
Fluctuation on precipitation	a greater ascent	18	1.8
	a slight ascent	236	23.2
	no obvious change	194	19.1
	a slight descent	353	34.7
	a greater descent	157	15.4

Other causes, including water resources, land, rainstorms, droughts, freezing hazards and population, also influenced the ecological environment of grasslands. Especially in the Heihe River Basin, the reduction of water resources and land, frequent droughts and freezing hazards, infrequent rainstorms and population growth were the natural and social causes for the ecological deterioration.

6.2.4 Evaluation on grassland management policies

1. Execution of grassland management policies

Grassland management policies could be classified into five categories, i.e. returning grazing into grasslands, returning farmlands to forests, grazing-related policies (including grazing-banning, grazing-off, and rotational grazing), livestock-forage balancing, and reward-compensation policies. Generally speaking, all these policies have been well implemented. Especially for ecological compensation reward policies, all households received government subsidies amounting to 14 259 yuan per annum on average, which was allocated on time and in full. Thus it can be seen that governments at all levels paid great attention on ecological reward and compensation, and a reward-compensation system had been established and feasibly enforced.

2. Implementation effect of grassland management policies

This survey statistically analyzed the implementation effects of grassland management policies (Table 6-4). According to the respondents, policies had been well implemented and had achieved certain effects in general. 37.5% of the respondents felt that overall effects of policy implementation were fairly good. The implementation of related policies had somehow controlled the inappropriate behaviors of farmers and herdsmen, such as crop farming over grasslands and damaging forests. Returning grazing into grasslands would reduce the quantity of livestock on pastoral areas, which would impose a positive externality on ecological protection of the Heihe River Basin. However, about 34.3% of respondents deemed that overall effects of policy implementation were mediocre.

Table 6-4 Effects of grassland management policy implementation in the Heihe River Basin

Effects of policy implementation	Options	Frequency	Relative frequency
Overall effects	Very good	151	14.8
	Fairly good	381	37.5
	Average	349	34.3
	Not good	26	2.6
	Fairly bad	9	0.9
Effects on income	a greater improvement	43	4.2
	a slight improvement	413	40.6
	no obvious impact	271	26.6
	a slight descent	162	15.9
	a greater descent	22	2.2
Effects on ecological environment	a greater improvement	91	8.9
	a slight improvement	624	61.4
	no obvious change	152	14.9
	a slight deteriorating	36	3.5
	a greater deteriorating	8	0.8

Regarding household incomes, 40.6% of the respondents felt that management policies did have a "slight" impact on income improvement. 26.6% of the respondents thought policy implementation had "no obvious impact" on household incomes. Therefore, grassland management policy implementation had

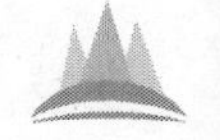

little impact on improving the household incomes of farmers and herdsmen. Even some respondents thought there were no impact on their incomes at all, or even imposed a negative impact on their incomes. Some respondents were not satisfied with their income in the recent years.

Regarding policy effects on ecological improvement, 61.4% of the respondents thought that there was "a slight improvement" on ecological environment compared with the situation a decade ago. Due to the reduction of grassland load capacity and the implementation of returning grazing into grasslands, the deteriorating trend of the ecological environment had stopped, but the improvement on the environment was fairly little. From the above analysis, management policies had little impact on improvement on the ecological environment as well as household incomes. There would be a long way to go to achieve the expected effects of policies.

3. Causal analysis on implementation effect of grassland management policies

Though most management policies have been well implemented, especially for the reward-compensation policies, here still exist other problems on management policies and the effects of those policies were not obvious. Both natural and social causes need to be discussed.

As to the natural reasons, according to Table 6-3, temperature increase and precipitation reduction were the main reasons behind the ambiguous effects on management policies. Furthermore, reduction of water and land resources, frequent droughts and freezing hazards, and infrequent rainstorm were the natural causes for the ecological deterioration in the Heihe River Basin.

Regarding social reasons, factors related to incomes and management were taken into consideration. Farmers and herdsmen tended to chase economic profits by destroying grasslands. If grassland administrators did not have a good sense of responsibility on management and punishment, less farmers and herdsmen would be punished for their violation. This would intensify the improper behaviors of violation. Conflicts and problems that would influence the effects of management policies could be summarized as follows.

1) Monotonous income source

As shown in Figure 6-3 and Table 6-5, major income sources of households in the Heihe River Basin were animal husbandry, farming, and migrant working, and the proportions of that were 22.9%, 22.4% and 9.6% respectively. Due to the severe natural environment, water shortage and arid climate, it is inappropriate to develop large scale of farming there, thus incomes from which would be limited. Also, migrant working was quite costly and made household departed from each other. Thus, most people were not inclined to do migrant working jobs. Therefore, animal husbandry became the major income source to support the local farmers and herdsmen. However, if the policy on returning grazing to grasslands was enforced and herdsmen were prohibited from grazing, their revenues would definitely been affected.

Subsidies alone could not solve the income problem of farmers and herdsmen. Since most of the farmers and herdsmen could hardly find other revenue-increasing means, they would get down to grazing again in order to sustain their own living, which would lead to grassland and the environment damage. This could be regarded as one of the reasons that governments hardly take actions or punishment on over-grazing and deteriorating on environment since the authorities could not find other feasible means to improve the income level for farmers and herds men in the short run.

Table 6-5 Major income sources of household income in the Heihe River Basin

Options	Frequency	Relative frequency
Animal husbandry	233	22.9
Crop farming	228	22.4
Livestock breeding	29	2.9
Ecological compensation	13	1.3
Trading	4	0.4
Migrating job–hunting	98	9.6

2) Lack of opportunities to increase income for the time being

The Table 6-6 shows that the ways for farmers and herdsmen to increase their revenues include sidelines (41.1%), migrant working (19.5%) and increasing grazing (11.3%). In the scenario of grazing banning, farmers and herdsmen could only choose sidelines as their top priority to increase revenues, as severe environment is not appropriate for farming while migrant working would cause family problems. Thus, sidelines, such as poultry-raising and cottage industries, would be the most realistic means to achieve income improvement. Expanding animal grazing against the government rule is usually their last resort to increase income, thus seldom facing severe punishment.

Table 6-6 Potential means to increase revenues in the Heihe River Basin

Options	Frequency	Relative frequency
Increasing grazing	115	11.3
Migrating job hunting	198	19.5
sidelines	418	41.1
Migrating to other regions	5	0.5
Keep unchanged	26	2.6

3) Insufficient supervision and punishment

When it comes to farmers and herdsmen violating the regulations on grassland management and protection (Table 6-7), 28.5% of the respondents deemed that they had "never" been punished while 25.6% of them had been punished "sometimes". It shows that administrative supervision and punishment in this regard is quite limited. Main reasons (with proportions) behind violating grassland management policies are listed as follows: chasing economic profits (26.2%), insufficient supervision (22.6%) and punishment (22%). Farmers and herdsmen were really eager to increase their revenues, but they had limited means and sources. Therefore, they chose to expand grazing scale, which led to the destruction of the environment. Since grassland administrators and governments could hardly find available revenue-increasing means, they would impose insufficient supervision and punishment on violations in this regard. This

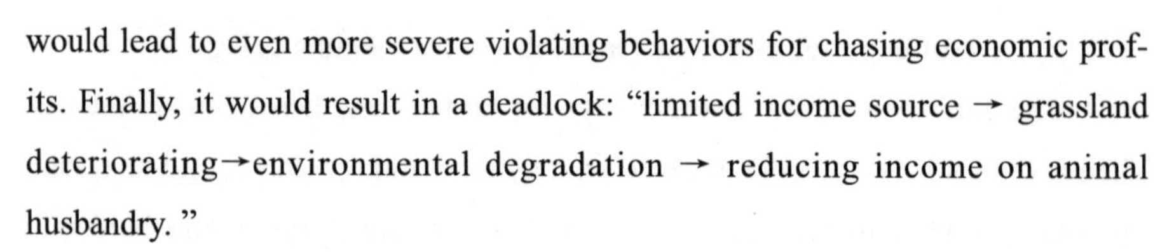

would lead to even more severe violating behaviors for chasing economic profits. Finally, it would result in a deadlock: "limited income source → grassland deteriorating→environmental degradation → reducing income on animal husbandry."

Therefore, the fundamental problem of grassland management in the Heihe River Basin is "monotonous income source". The underlying conflict that needs to be urgently solved is the conflict between the monotonous income source and the ecological protection of grasslands. Based upon that, further grassland management policies should be supplemented by measures regarding income improvement as the essential guarantee for grassland ecological protection. Supporting mechanism would tackle both the cause and the effect of the conflicts, which would finally lead to the good governance and improvement of grassland environment.

Table 6-7 Violations of grassland management rules

Questions	Options	Frequency	Relative frequency
Have you ever been punished due to the violation?	Frequent	15	1.5
	Sometimes	260	25.6
	Average	131	12.9
	Occasionally	214	21.0
	Never	290	28.5
Main reasons for violation?	Chasing economic profit	266	26.2
	Insufficient supervision	230	22.6
	Insufficient punishment	224	22.0
	Imitating others' violation	61	6.0
	Others	42	4.2

6.3 Policy Recommendation for Grassland Management

6.3.1 The core of the administrative policy for the government

In order to solve the problem of "monotonous income source", diversified

policy systems on increasing revenues would be the core of the administrative policy for the government.

On one hand, the key issue on the current agenda is to "increase the revenues for farmers and herdsmen". The means of subsidizing farmers and herdsmen, enforcement on returning farmland to grassland and returning grazing to grassland could not radically solve the environmental problems. To solve the problem of "monotonous income source" would be fatal to increase the revenues for farmers and herdsmen in the long run. By diversifying their means to obtain economic profits, partly migrating population, and increasing employment opportunities would be the basic measures to solve the income problem. The basic problem of management on grazing lies in the current policies themselves. Subsidy and grazing-banning could not solve this basic problem. Being independent from the grassland-oriented revenue-increasing means would be the practical measures to prevent the violating acts of locals and grassland damage. Breaking the deadlock of "limited income source → grassland deteriorating → severe environment → reducing income on animal husbandry" would be the foundation of the ecological protection as well as the core of the administrative policy for the government in the case of the Heihe River Basin.

On the other hand, the core of the administrative policy for the government is to "establish the diversified income source structure". Governments should guide and support locals to develop sidelines as well as the tertiary sectors, which would be beneficial to extending the income source. The problem could be solved through introducing emerging industries to diversify employment opportunities and fully utilize surplus labor force. Therefore, amending and improving grassland protection policies would still be the key issue for the further administration. Policy systems on increasing revenue for farmers and herdsmen would be beneficial to the protection of living environment of grassland and the settlement of the crux of grassland management.

6.3.2 Policy recommendations on grassland management policy-making for the Heihe River Basin

1. The foundation of grassland management in the Heihe River Basin: Enhancing administrative supervision through amending and improving grassland protection policies

Regarding the key points of further policy-making of grassland management, 42% of the respondents thought that it is necessary to modify the current policies. Based upon the policy amending and improvement, sufficient supervision (22.6%) as well as publicity and training (10%) were also essential (Table 6-8). The enforcement of current policies was not sufficient to protect grazing areas, forests and grasslands. More strict measures should be taken by means of supervision, punishment and legislative publicity. Based upon improved policies and sufficient supervision, the ecological environment could be protected and improved.

Table 6-8 Emphases on further policy-making of grassland management for the Heihe River Basin

Options	Frequency	Relative frequency
Sufficient administrative supervision	230	22.6
Amending or improving ecological protection policies	425	41.8
Publicity and educational training	102	10.0

2. The core of the grassland management policies in the Heihe River Basin is to increase the revenues of local farmers and herdsmen

The overall income level and the living standards of the farmers and herdsmen in the Heihe River Basin are not high, and need to be improved. Governments should help farmers to bring in emerging industries or industries of the tertiary sectors to extend the income source of the locals; introducing emerging industries would diversify the employment opportunities and fully utilize surplus labor force, which will be helpful for solving the employment problem. Policy systems to diversify the income sources for farmers and

herdsmen would be beneficial to breaking the current deadlock. In addition, it would be important to overcome the conflict between monotonous income source and ecological protection of grassland.

Industrial structure transformation is one of the important ways to increase the revenues for every household. Animal husbandry is the basic industry in the Heihe River Basin, which should be better developed to enhance the income level. It is necessary to cultivate a stable and sustainable sales market and open new channels for livestock products. Further by-product processing should be encouraged and supported by governments. As to breeding mode, standardized breeding should be advocated by barn feeding to reduce grazing. Other cash crops could be introduced and cultivated, such as desert cistanche. Financial supports could also be needed to encourage farmers to be self-employed.

3. Other issues deserving government's attention for further policy-making

(1) Adjustment in reward-compensation policies. It is necessary to increase grassland compensation, and raise the subsidy for livestock-forage balance and artificial pasture recovery, which would be an important aspect to adjust reward-compensation policies. Meanwhile, the government should increase subsidies for the livestock breeding, agricultural machinery, water resource usage, grassland attendants, and minority students. The living and production conditions would be improved through raising subsidy standards and enlarging subsidy scope.

(2) Strengthening environmental governance. Because of deteriorating environment, grasslands are facing such problems as resource shortage, grassland pasture degradation, and gradual arid climate, which call for environmental governance and policy-making on grassland protection. Grassland management and recovery should be arranged at the levels of county and village by raising the reward and compensation levels. Efforts should be reinforced to control the plague of insects (grasshoppers) and protect water resources.

(3) Promoting infrastructure construction. Governments have the responsibility to increase the investment into infrastructure construction, especially those for drinking water supply, water dams and sensible exploitation of underground water resource, which would be the crux to the herdsmen's water security. Transportation construction is also necessary. Hydroelectric station construction would guarantee the power usage of daily life. In general, the rural environment, rural planning and health conditions should also be improved.

(4) Strengthening social security. Grassland policy-making should consider the system of registered permanent residence to ensure the basic living of farmers and herdsmen. As to the social security, attention should be paid on the following details: to overcome the educational difficulties for the children of farmers and herdsmen since currently educational quality is pretty low and the cost is pretty high; to improve the medical care standard and overcome the difficulty of hospitalization; more employment opportunities should be offered to herdsmen's children; to relocate and resettle the housing of herdsmen who have devoted to grazing-banning and grazing-off; and to strengthen the pension security system.

Chapter 7

An Analysis of Grassland Management Policies Based on an Opinion Survey of Grassland Administrators

7.1 Questionnaire Design

7.1.1 Questionnaire content

The research team also designed a questionnaire for grassland administrators. It aimed at understanding the current situation of policy implementation according to the feedback from related administrators. First-hand information has been collected to support the related research of this project. An analysis of the survey results from farmers, herdsmen and administrators was carried out as well.

The questionnaire was drafted and designed based upon in-depth discussions among research members and suggestions from experts in the fields. Content of the questionnaire comprise three sections, including the basic information of respondents, ecological environment and policy implementation, and effects of policy implementation. Its questions are summarized in Table 7-1.

Table 7-1 Summary of main contents of the questionnaire

Aspects	Content
Basic information of respondents	Basic information of respondents: nationality, gender, age, educational level, administrative rank, length of employment Household characteristics: household size and number of working-age people, and quantity of durable goods
Ecological environment & policy implementation	Ecological environment: temperature change, precipitation change, and other factors Evaluation on grassland ecological protection policies Reward-compensation policies on grassland ecological protection
Effects on policy implementation	Governments' attitudes towards grassland protection Local supervision on grassland protection policies or measures Punishment on behaviors damaging grasslands Reasons for violating grassland management policies Impact of grassland protection policies or measures on incomes Overall governance effects of grassland protection policies or measures Key issues on future grassland management policy-making

The first section concerns questions about basic information of respondents, including nationality, gender, age, administrative rank, educational level, and political status, length of employment and categories of departments he or she serves. All the above information would be sufficient to ensure the authenticity and validity of survey results. During the survey, copies of the questionnaire were handed out according to the weightings of nationality, gender, administrative rank and categories of departments. Older respondents with higher education background and longer service years were selected for the survey.

The second section includes questions about the current situation of ecological environment and the natural causes of the ecological and environmental alteration, and management policies implemented in recent years.

The third section was designed to find out the implementation effects of current policies and their causes, and the supervision and punishment carried out by administrators. Survey results are significant to analyze the effects of policy implementation, and to compare with the survey results from farmers and herdsmen. Conclusions are drawn from the above data and result analyses.

7.1.2 Survey plan

To ensure the validity of its results, the research took field survey as the main investigation mode. Researchers and team members from the Institute

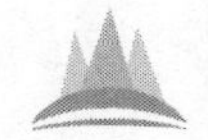

of Policy and Management, CAS and Lanzhou University made field trips throughout related counties and areas. In face-to-face discussions and interviews, first-hand data and information were collected from administrators at grass-root, municipal, and provincial levels. To obtain large enough samples, the field surveys covered all the counties in the Heihe River Basin. Of the 250 copies of questionnaire issued, a total of 233 copies of valid responding questionnaire were received (183 from Gansu Province, 42 from Inner Mongolia Autonomous Region, and 8 from Qinghai Province). The rate of validity reached up to 93.2%. Distribution of the questionnaire among counties and related areas was classified according to the provincial and municipal units and other related administrative organizations. Geographically speaking, 42 respondents from Ejina Banner in Inner Mongolia Autonomous Region, 8 respondents from Qilian County in Qinghai Province, 2 respondents from Lanzhou City in Gansu Province, 157 respondents from Zhangye City, 24 respondents from Jinta County in Jiuquan City. In Zhangye City, 46 respondents were from municipal units, with 4 respondents from Ganzhou District, 30 respondents from Gaotai County, 18 respondents from Linze County, 44 respondents from Minle County, 15 respondents from Sunan County. Relatively speaking, the questionnaire was equally distributed in all the related counties, cities and provinces which would authentically reflect the current situation.

7.2 Survey Results

7.2.1 Characteristics of respondents

1. Basic characteristics of the respondents

The respondents of this survey were administrators from related areas and organizations. Most of them worked at departments in charge of the supervision and legislation on grasslands, animal husbandry management, technology promotion, or other specific tasks on administrative management. Regarding their ethnic characteristics, the majority of them were Han nationality (76.8%) while a substantial proportion of them were Mongolians and Yugurs. For the respondents' gender characteristics, more than 70% of them were male. As to their age distribution, 39.1% were in the age range of 41-50. According to political status, more than 60% of them were CPC members. About 63.1% of their administrative ranks were at the minor level and below. That is, most of the local interviewees were experienced administrators in charge of the pastoral production and grassland management. In terms of educational levels, the majority of them (70%) were trained in universities (or junior colleges and above), showing that they were well educated and had a better understanding of the current situation and problems on pastoral production and grassland management. As to the length of employment, more than 50% of them worked for more than 10 years, meaning more than half of the respondents were quite experienced in their jobs and had a comprehensive understanding of the current situation and policy implementation. This guaranteed the authenticity and validity of this survey. According to the statistical characteristics, the respondents were rationally distributed and their answers significantly reflected the real situation of the problems.

2. Basic characteristics of households

The average household size of the respondents (Figure 7-1) was 3.5 people. There were 2.2 working-age people in a household on average. Of the working-age people in each household, 1.88 were devoted to agricultural production and 0.73 to animal husbandry. The proportion of those working in agricultural job is much higher than those in grazing. That is to say, most administrators in this survey were from farming areas, and a few were from pastoral regions, showing the current situation that more households are devoted to the farming[①].

① It can also be explained that family members of local administrators were more inclined to agricultural production than pastoral work.

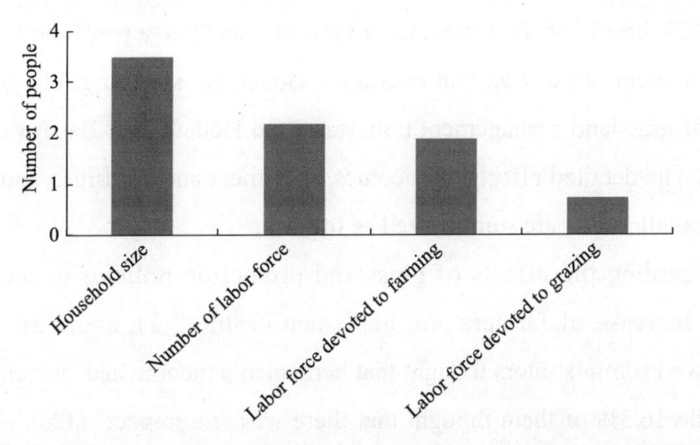

Figure 7-1 Basic characteristics of households of administrators in the Heihe River Basin

7.2.2 Survey on grassland management policies

1. Implementation of grassland management policies

This survey concerns grassland management policies in five aspects: returning grazing to grassland; returning farming to forestation; rotational grazing, grazing-off, grazing-banning; livestock-forage balance; and ecological reward-compensation. According to the administrators in the survey, the ranking of the importance of the above five polices may be listed as follows: ecological reward-compensation policies > policies on rotational grazing, grazing-off and grazing-banning > policies on returning farming to forestation > policies on returning grazing to grassland > policies on livestock-forage balance. All the policies were feasibly implemented, especially for grassland ecological reward-compensation policies. From the administrators' points of view, governments at all levels have always paid "relatively high attention" to grassland protection, especially the implementation of grassland ecological protection reward and compensation since the reward-compensation mechanism on grassland ecological protection was feasibly established.

2. Policy execution effects

As to the overall governance effects of the grassland protection polices

(Table 7-2), 46.4% of the respondents viewed it as "fairly well", however, about 24.9% of them rated it as "on average". Generally speaking, the governance effects of grassland management policies in the Heihe River Basin were just on average. The detailed effects on incomes of farmers and herdsmen and grassland ecological alteration are summarized as follows.

Regarding the effects of grassland protection policies or measures on income increase of farmers and herdsmen (Table 7-2), about 36.1% of the interviewed administrators thought that herdsmen's income had "increased a little bit" while 16.3% of them thought that there was "no impact" at all. Even about 18.9% of the administrators thought there was "a slight decrease" in the incomes of farmers and herdsmen.

Table 7-2 Effects on grassland policy implementation of administrators

Policy effects	Options	Frequency	Relative frequency
Overall governance effects	Very well	16	6.9
	Fairly well	108	46.4
	On average	58	24.9
	Fairly poor	4	1.7
	Very poor	1	.4
Effects on income	Improved a lot	18	7.7
	Improved a bit	84	36.1
	No impact	38	16.3
	Decreased a bit	44	18.9
	Decreased a lot	5	2.1
Effects on ecological environment	Improved a lot	36	15.5
	Improved a bit	95	40.8
	No impact	12	5.2
	Decreased a bit	32	13.7
	Decreased a lot	19	8.2
Current situation of ecological environment	Decreased grassland area	40	17.2
	Decreased grassland species	40	13.7
	Increased inferior species	15	6.4
	Soil deterioration	30	12.9

In terms of the administrators' perceptions on grassland ecological

environment (Table 7-2), 41% of them thought that the ecological condition of grasslands had "improved a little bit" in the recent decade, and about 13.7% of the interviewees thought that the grassland environment had "deteriorated a little bit". Generally speaking, ecological deterioration would include the following three aspects: grassland area decreasing (17.2%), less grassland species (13.7%), and soil deterioration (12.9%).

Based upon the above analysis, conclusions are summarized as follows. Policies on grassland management and protection had some effects, including overall governance effects, environmental improvement and effects on income. However, they had limited impact on environmental improvement and income enhancement. Overall effects of governance policies on grassland environment were pretty general.

3. Causal analysis of policy effects

There are problems with grassland management policies in the Heihe River Basin. The reasons may fall into two aspects: natural factors and man-made factors (Table 7-3).

The natural factors causing ecological deterioration include: decreasing precipitation, rising temperature, decreasing water resources and land resources, increasing droughts, decreasing rainstorms, and increasing freezing hazards. Mostly, temperature and precipitation are the two major factors causing ecological deterioration. Statistically speaking, 54.5% of the respondents regarded decreasing precipitation as the main factor while 51.5% of the respondents regarded increasing temperature as the main natural factor. Meanwhile, other natural factors also influenced the ecological condition of grasslands in a direct and indirect way, which were also significant in causing local grassland ecological alteration.

Man-made factors are analyzed into the following two angles: herdsmen and administrators. Insufficient punishment (22.3%) and insufficient management (12%) are two main administrative problems; inclination and motivation on chasing profit and income improvement (18.9%) would be the two main problems on the farmers' side.

Table 7-3 Causal analysis on effects of policy implementation according to administrators in the Heihe River Basin

Man-made factors	Options	Frequency	Relative frequency
Reasons for policy violation	Income improvement	44	18.9
	Insufficient management	28	12.0
	Insufficient punishment	52	22.3
	Imitating others' violation	11	4.7
	Others	27	11.6
Extent of supervision	Very sufficient	25	10.7
	Fairly sufficient	91	39.1
	On average	59	25.3
	Fairly insufficient	12	5.2
	Very insufficient	4	1.7
Extent of punishment	A lot	10	4.3
	A little	99	42.5
	Not much	61	26.2
	never	21	9.0

Statistically speaking, regarding the extent of supervision, 39.1% of the administrators thought that the extent of supervision was "fairly sufficient", 25.3% of them thought that was "on average", and 10.7% of them thought that was very sufficient. From that point of view, administrative supervision on grassland management and protection policies was quite general. As to farmers' and herdsmen's violation on grassland management policies, 42.5% of violations were punished "lightly" and many violations were unpunished at all. Therefore, the extent of punishment on local violation is quite limited.

According to the comprehensive analysis, local governments paid considerable attention to grassland protection. However, there were still some violations of rules, which mitigated the governance effects. The main reason behind the current problem is the insufficient punishment. Therefore, more severe punishment should be taken against the violations. Meanwhile, more supplementary measures should be introduced to improve the incomes of farmers

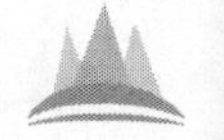

and herdsmen and to alleviate their economic–chasing behaviors, such as husbandry production with grassland deterioration.

7.2.3 Government priorities on grassland management

According to administrators, the priorities of future grassland management should include the following three aspects: modifying grassland protection policies, strengthening the implementation of current policies, and improving the herdsmen's income.

First of all, grassland protection policies should be modified and improved. The current grassland management policies should focus upon the core management problem, which is to improve the revenues of farmers and herdsmen. Based upon that, strengthening the policy implementation and the publicity of grassland protection would finally enhance the overall governance effects of policy implementation.

Secondly, insufficient policy implementation, punishment and supervision would be the main factors causing anthropogenic damage to grasslands. Therefore grassland supervision should be enforced, including supervisory team development, sufficient increase in capital, personnel and facilities, and toughening penalties for rule violation, which would finally strengthen the grassland supervision and management. Also, other policies should be strengthened, including the continuity of policy implementation, increasing the financial support to reward-compensation system, improving employees' occupation accomplishments, and increasing the coverage and publicity of grassland protection policies. Meanwhile, the interest of administrative personnel (including their social security and treatment) should also be effectively secured.

Moreover, it is necessary to improve the income level of herdsmen and farmers. Actually farmers and herdsmen had a clear awareness on grassland protection. However, due to their limited revenues, they had no other means to improve their income level but overexploiting grasslands. Therefore, further policy-making on grassland management should focus upon the interest of farmers and herdsmen. Improving their income level as well as living standards would be the key to ameliorate the grassland ecological environment. Meanwhile, social security on drinking water, housing, education and medical care should be sufficient to secure the living and production for farmers and herdsmen. Transforming the unfeasible production mode and lifestyle would be necessary for grassland protection with more stocking breeding but less grazing.

Finally, it is essential to strengthen the environmental governance by alleviating environmental and agricultural pollution. Efforts to control the impact of chemical fertilizers on the environment and soil, and pest management would also be important. Overall environmental governance in the Heihe River Basin would be realized through comprehensive strategies to address both natural and man-made factors.

7.3 A Comparison of the Survey Results

7.3.1 Result comparison

1. Regarding the effects of grassland management policies

First of all, a comparison of the responses from the two groups (farmers and herdsmen on one side, and administrators on the other) shows that they have similar opinions on the policy implementation. According to the respondents, policies in the five categories have been well implemented, which means that governments at all levels have devoted much effort to the implementation of management policies, especially for reward and compensation system on grassland ecological protection. Thus, in general, policies on reward-compensation and financial support to achieve grassland management have been well implemented in the western areas of China.

Secondly, when comparing opinions on the policy effects, we also find that the two groups of respondents come to the similar conclusion. Farmers, herdsmen and administrators thought that policies on grassland management had exerted

quite an impact on controlling the violation behaviors, such as deforestation, inappropriate grazing and farming on the pastoral areas. Management policies, particularly in returning grazing to grassland, had produced some effects on reducing pasture's loading capacity of livestock, which would be of great significance to the recovery of grassland ecological environment in the Heihe River Basin. It reveals that effects or good performance of the current management policies had been witnessed by local farmers, herdsmen and administrators. Additionally, it demonstrates that the ecological environment of the western areas has been improved significantly through the efforts of the governments at all levels as well as the financial supports to the ecological protection of grassland, forests and green lands.

Nevertheless, the two groups thought that the overall effects of the policies did not perform well. A large number of respondents thought that the governance effects of grassland management policies were just "on average". When asking about the effects of the policies on environmental improvement and income, most respondents thought that most policies had not reached the expected performance. Considering their effects on income improvement, most respondents thought that the current policies had no impact on the improvement of farmers and herdsmen's incomes, and even some respondents thought that their incomes had been leveled down due to the policy implementation. Similarly, a substantial number of respondents thought that grassland management policies had no, or even worse, impact on the improvement of living and ecological environment of the Heihe River Basin. Therefore, grassland management policies, on general, were quite significant especially for restraining continuous worsening ecological environment. However, the improvement on grassland environment and living standard for farmers and herdsmen were not obvious. It means that there are problems with the current policies themselves, which needs to be modified on the procedure of further policy implementation.

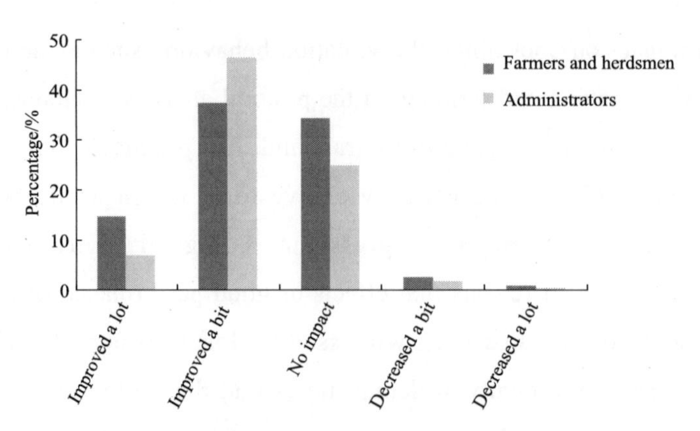

Figure 7-2 Comparison on responses among farmers, herdsmen and administrators on overall governance effects on management policies

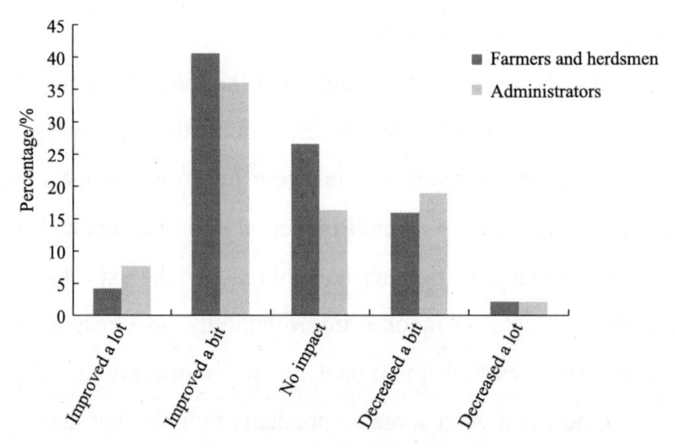

Figure 7-3 Comparison on responses among farmers, herdsmen and administrators on income effects due to policy implementation

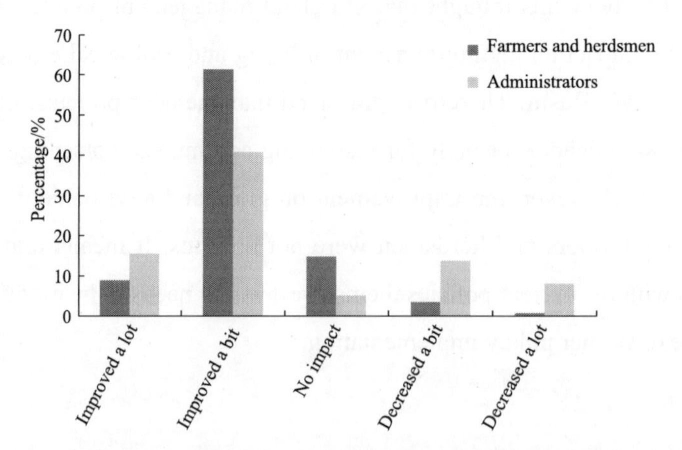

Figure 7-4 Comparison on responses among farmers, herdsmen and administrators on effects on ecological environment due to policy implementation

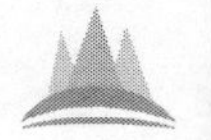

2. Causal analysis on results of grassland management policies

Temperature and precipitation would be the two main natural causes. Rising temperature and decreasing precipitation would be the major causes that made it difficult to improve the ecological environment of the Heihe River Basin in the recent decade. Besides, decreasing water resources and land resources, increasing droughts and freezing hazards, and decreasing rainstorms would also be the possible cause for the environmental degradation.

Regarding man-made factors, farmers, herdsmen and administrators generally agreed that three causes resulted in the ecological deterioration of grasslands in the Heihe River Basin: insufficient punishment, inclination on profit-chasing and insufficient management. However, preferences of farmers and herdsmen were different from that of administrators when reviewing the current policies. From the angle of administrators, they thought that "insufficient punishment" would be the main factor causing the violation of farmers and herdsmen rather than "insufficient management". Accordingly, by strengthening the extent of punishment would effectively alleviate the violation behaviors. However, the survey results from farmers and herdsmen showed that most of the respondents were concerned more upon their own profits. Due to their monotonous income source, all they could get rid of poverty and raise their income level is overgrazing and pastoral farming, which would be the behaviors resulted in the grassland deterioration.

In summary, insufficient punishment, inclination on profit-chasing and insufficient management would be the three main reasons causing the violations. This conclusion had been reflected on the responses of farmers, herdsmen and administrators. This is also the primary reason behind the average (or even worse) performance of grassland management policies in the Heihe River Basin.

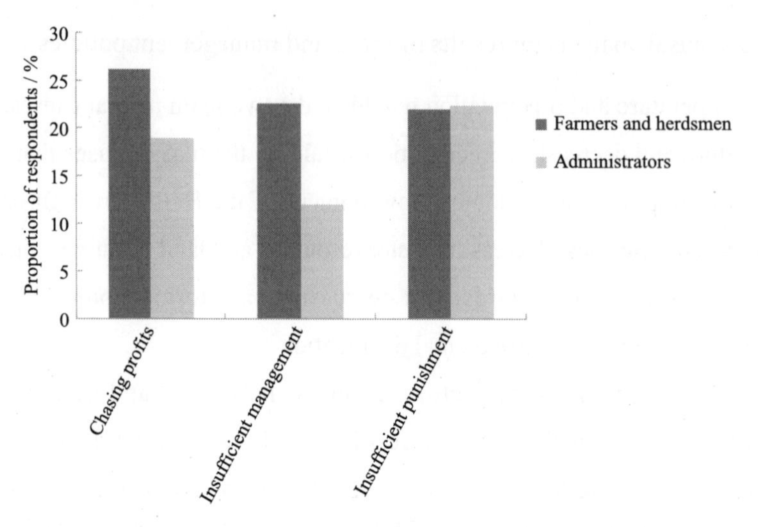

Figure 7-5 Comparison on responses among farmers, herdsmen and administrators on grassland violation reasons

7.3.2 Conclusion of comparisons

Final conclusion would be drawn from the above comparisons. Responses from farmers, herdsmen and administrators all come to the same conclusion when considering the main reasons for the violation of grassland regulations: insufficient punishment, inclination on profit-chasing and insufficient management, which lead to grassland degradation and the limited performance of management policies.

Policy-making that ignores the income issue of farmers and herdsmen would be the crux of the current policies, which causes frequent violations. Administrative countermeasures only emphasize ecological management and protection rather than the improvement of living standards and income level. Ecological problems in the Heihe River Basin could not be fundamentally solved solely by reward-compensation policies and monotonous punishment measures. All those measures could only treat the symptoms rather than the root cause which would finally lead to the limited performance of the current policies. A deadlock of "limited income source → grassland deteriorating → environmental degradation → reducing income on animal husbandry" would come into the end. Therefore, the

final solution to overcome the related issue is to diversify the income source of farmers and herdsmen and try to raise their income level. Other comprehensive measures should be supported to realize and strengthen the general governance on ecological protection of grasslands.

Generally speaking, solutions to overcome the ecological protection issue in the Heihe River Basin include: bringing in emerging industries or the tertiary sectors for the locals to extend the income source, introducing some emerging industries that would diversify the employment opportunities and fully utilized surplus labor force, which will solve the employment problem. Policy systems to diversify the income source for farmers and herdsmen would be beneficial to changing the current deadlock, meantime, it would be sufficient to overcome the conflict between monotonous income source and ecological protection of grasslands. Environmental governance from the perspective of natural factors and the improvement on income level, sufficient management and punishment from the perspective of man-made factors would be effective to establish the comprehensive supporting policy system. This would finally lead to the improvement of grassland ecological environment as well as living standards of local farmers and herdsmen in the Heihe River Basin.

Chapter 8

Administrative Problems Challenging Sustainable Development in Western Grasslands and Countermeasures

8.1 Problems Challenging Sustainable Development in Western Grasslands

8.1.1 Grassland protection and herdsmen's income

Since the implementation of grassland ecological protection policies, the exploitation and utilization of grassland resources has been limited and so have been the development of husbandry and the growth of herdsmen's revenues. This has also constrained the social economic development of pastoral regions. According to the field survey of this book, the household income of herdsmen and farmers in the Heihe River Basin was comprised of three parts: revenues from husbandry, agricultural farming and migrant working. The average annual household revenues from these three sources were 33 310 yuan, 24 968 yuan, 20 686 yuan respectively. With regard to annual net incomes, three quarters of the households ranged

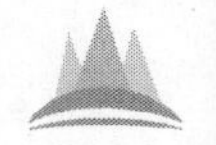

from 10 000 to 70 000 yuan while more than 30% of them ranged between 30 000 to 50 000 yuan. Most of the responding households were low and middle-income families. Accordingly, animal husbandry was their major income source, therefore, grassland protection policies, such as grazing-off and grazing-banning, exerted an apparent impact on local household incomes.

Confronting such a high opportunity cost, herdsmen would not be willing to protect ecological environment of grasslands. "Grazing illegally in secret and at night" happened frequently and was difficult to prohibit. According to the *Report on Statistics and Analysis of the National Grassland Illegal Cases in 2014*, the total number of grassland illegal incidences were 18 998, of which 17 848 were put on file with the filing rate up to 93.9%. A total of 17 423 cases were settled with the settlement rate up to 97.6%. A total of 20 900 hectares of grasslands were damaged due to illegal exploitation, expropriation and temporary occupation. A total of 1220 hectares of grasslands were traded or exchanged illegally. A total of 621 cases were transferred to judicial departments, and 25 were filed for administrative reconsideration or administrative proceeding. In 2014, there were 14 899 cases involving the violation of the rules of grazing-off and grazing-banning, accounting for 78.4% of all the illegal cases on grasslands, ranking first among all kinds of violations (Figure 8-1).

Illegal grassland exploitation ranked the second (10.1%) with the total number up to 1910. That was followed by the violations on provisions of livestock-forage balance with the total of 1013 cases.

This proves that the principal problem facing grassland management is to improve the incomes of farmers and herdsmen. The continuous improvement of grassland ecological environment and economic development in pastoral areas could be achieved by extending financial and income channels for farmers and herdsmen and raising their self-consciousness on grassland protection.

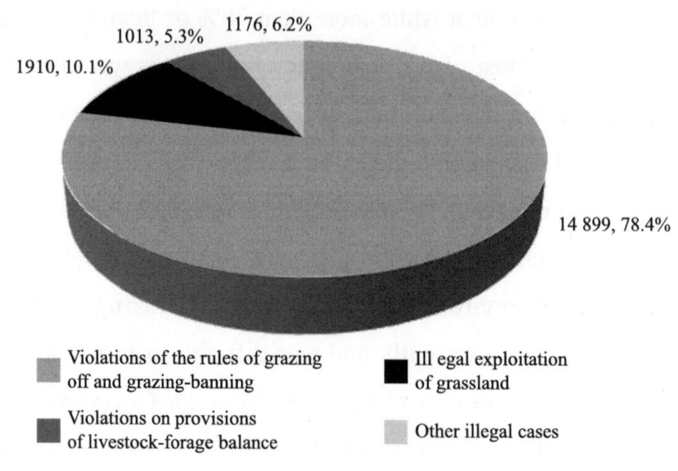

Figure 8-1 Compositions of different kinds of grassland illegal cases

8.1.2 Inconsistence and uncertainty of protection policies and herdsmen's living and production in the long run

Grassland ecological environment issues have drawn much attention from all walks of life in recent decades. Besides searching for the root causes of ecological degradation, the impact of grassland degradation on herdsmen's living and production has also been investigated and analyzed. Related studies show that over-exploitation and unplanned migration would cause severe deterioration and desertification (Bao and Zhou, 2001). Some studies reveal that mistaken policy orientation, irrational exploitation of grassland resources, and extensive mode of operation resulted in ecological degradation (La, 2005). Due to the severe imbalance of grassland eco-system, large areas of desertification and degradation worsened the living conditions of local herdsmen and produced poverty. Apparently, grassland ecological degradation is not the unique factor influencing herdsmen's living and production.

Since the 1990s, the initiative to contract the managerial rights of grasslands was expected to be a means to avoid "tragedy of the commons". However, contracting operation had not reached the expected consequence, and this form of operation had destroyed the integrity of grasslands and interrupted the

local regulations, which finally led to the conflicts on grassland utilization and exploitation. After the year of 2000, subsidies and other supporting measures were introduced to adjust the micro-scale production behaviors of herdsmen. However, ecological environment had not been improved in most of the regions due to the widespread illegal behaviors and violations. Failures of policy intervention at the national level have revealed that oversimplifying decision-making and re-centralization are the primary problems that should be overcome in the near future.

8.1.3 Herdsmen's living costs and the extent of compensation policies

Due to the policy implementation on grazing-banning and livestock-forage balance, herdsmen found it difficult to raise their revenues while their infrastructure costs and living expenses were going up. According to the field survey and responses from farmers and herdsmen, future grassland management should place a high priority on "modifying and improving grassland protection policies", besides, strengthening supervision and publicity on policy implementation are also important. However, problems facing grassland management in the Heihe River Basin were mainly in the policy itself, rather than the lack of awareness on grassland protection of herdsmen and farmers. Somehow profit-chasing is not their intentional behaviors. Therefore, it is necessary to strengthen the supervision on policy-implementation while modifying and improving grassland management and protection policies. This is the primary cause to tackle the grassland management dilemma in the Heihe River Basin.

According to the field survey on policy implementation, statistical analysis on government subsidy distribution was carried out. On average, governments provide each household with 14 259 yuan each year with the maximum up to about 210 000 yuan. Most of fiscal subsidy could be distributed timely and the proportion of that accounted for 64.6%. It could also be fully funded and the proportion of that accounted for 67.7%. Regarding its impact on household income, about 40.6% respondents thought that related policy implementation did improve the household income. 26.6% respondents thought there was no impact on household income at all and even about 16% respondents reviewed that there was a substantial decline on household income due to grassland protection policies.

Therefore, grassland management could not only be achieved by subsiding farmers and herdsmen or grazing-banning. The key issue on grassland management for the Heihe River Basin is to "increase farmers and herdsmen's revenues". The core of policy-making and basis on ecological protection in the Heihe River Basin is to overcome the problems of monotonous income sources and improve the income level of local farmers.

8.2 Countermeasures on Promoting Management Ability for Sustainable Development of Western Grasslands

8.2.1 Grassland property right clarification as a basis for management

It is important to clarify the ownership and rights to use grasslands, because they are the basis for the grassland household contract responsibility system. The connotation, scope and execution of the rights need to be further specified. It is also important to have normative legal and institutional guarantee, and the codes of practice. A clear specification and definition on the rights of use, usufruct and disposition would facilitate merchandizing the rights and permit contracted grasslands to be leased freely in the form of joint-stock cooperation. To establish a system to determine the carrying capacity of grasslands, violations of the codes or provisions on grassland carrying capacity would be charged with grassland use and overloading fees. Those who have caused severe desertification and not

willing to rectify their violations should be legally punished. Their contracted grasslands should be withdrawn and the contract terminated. In the principle of "making those who cause destruction to pay for it", governments should see to it that vegetation is restored on pastures damaged by mining or hydraulic engineering projects.

Grassland contractual management should be beneficial both to ecological protection and revenue-increasing for farmers and herdsmen. It is essential to adopt measures suiting local conditions by introducing diversified management measures in different regions. Grasslands in farming-pastoral regions are the means for farmers and herdsmen to develop sideline production. Giving equal emphasis on agriculture and animal husbandry is the main character of the region and moderate household contract is beneficial to the grassland utilization. Most farmland on grassland is not contracted to households and the area per capita is pretty small. Local people's self-consciousness on grassland protection needs to be raised. Specification on grassland attributes through bidding contract is a realistic method.

8.2.2 Enhancing herdsmen's participation in grassland management to strengthen the development of grassland communities

Participatory grassland management planning should be carried out on the basis of communities (or villages). With the joint participation of administrator, technicians, farmers and herdsmen, especially the involvement of the community members and stakeholders, the grassland management planning should be based on the need evaluation at various levels, and suggestions and advices from all walks of life.

Through training programs at various levels, the capacity of technical and managerial professionals at the district and township levels will be developed and the awareness of herdsmen and farmers to participate into grassland management enhanced. This participative management will arouse the enthusiasm of herdsmen for daily grassland management and environment protection. It would facilitate the changes in herdsmen's behaviors and customs, which would be conducive to the recovery of grassland production, the mitigation of grassland degradation, and the improvement of the living conditions of herdsmen.

It is advisable to encourage the establishment of grassland cooperatives between herdsmen and non-herdsmen so as to take full advantage of the latter's finance, technology and social connections. It is necessary to do a good job in providing herdsmen with pre-productive and post-productive services, so as to improve the profits of livestock products, raise the incomes of herdsmen, and promote the economic development in pastoral areas.

8.2.3 Optimizing strategic layout of pastoral management and enforcing targeted measures

It is advisable to effectively optimize the industrial structure of pastoral regions. It is important to give full play to animal husbandry, the pillar and competitive industry in the pastoral regions. It is significant to energetically develop the processing industry with livestock products as raw materials and other secondary and tertiary industries in the regions through the introduction of advanced technologies and financial support. By adjusting industrial structure, population pressure in eco-vulnerable pastoral areas could be eased by transferring labor force to secondary and tertiary industries.

Eco-migration policy should be resolutely implemented in areas suffering from severe ecological deterioration. Because of severe desertification and fragile ecology, living conditions have been lost in some pasture regions, where residences are scattered and the costs for education, transportation and medical care are very high. Through eco-migration, herdsmen and farmers in eco-vulnerable areas would be collectively relocated to places with better living conditions. This may provide opportunities for ecological vulnerable areas to recover and facilitate the fundamental solution to ecological and poverty difficulties by changing living and production patterns of herdsmen.

It is advisable to vigorously expand ecological tourism in pastoral areas.

Thanks to their unique characteristic, grassland areas enjoy unique climate and vegetation conditions, and the characteristics of eco-systems, social and humanistic background, which would be the favorable endowment to cultivate eco-tourism. By rational exploitation of tourism resources, the population and economic endowment in pastoral areas would be transferred to the service sectors. Furthermore, it is a feasible means to expand local income source and employment.

8.2.4 Establishing an effective system of grassland management and a long-term regulatory regime

It is important to tighten the supervision of grassland resource exploiting industries, such as mining, sand pits and tourist attractions by rationalizing the functions and power of administrative organs at different levels, and making strict approval procedures for grassland requisition. It is also important to do away with activities damaging grasslands, such as illegal and irrational exploitation, and punish rule-violating behaviors, including malicious destruction of grassland ecology and alteration of grassland usage.

The grassland ecological monitoring and precautionary system needs to be improved by establishing various monitoring stations in line with different species and distribution of grassland resources, and according to the technical specification of ecological monitoring. Ground positioning and monitoring stations need to be enhanced by amending and modifying the related technical codes and practices with regard to field survey and remote-sensing data analysis. Efforts should be made to integrate advantageous techniques and equipment housed in universities, research institutes and technology promotion centers so as to speed up the transformation and application of new techniques and research achievements in grassland monitoring.

It is important to strengthen fire precautionary fighting work in pastoral areas by establishing closer cooperation with meteorological and fire prevention branches. Administrative departments at all levels should be responsible for convening fire-precautionary meetings regularly to implement the duties and responsibilities, and preparing emergency agendas. Strictly self-checking and selective checking are necessary for fire precautionary supervision, in particular during seasons of spring and autumn. Training on technical maintenance on application of fire precautionary information system needs to be carried out and the equipment of fire precaution needs to be reserved and utilized. Emergency management on grassland snow hazards would also be taken into consideration by formulating emergency plans.

To promote grassland management capacity, it is important to enhance academic research, and promote cooperation and communications with counterparts in the field both at home and abroad through resource and information sharing.

8.2.5 Promoting the development of cooperatives and forming an administrative pattern featuring multi-participation

Specialized cooperatives of farmers and herdsmen would facilitate the development of both rural productive forces and local economy. Therefore, it is necessary for governments to set up a more feasible environment in favor of cultivating specialized cooperatives. It is necessary to bring in leading enterprises, especially those in agricultural production, and enhance their ability to drive the cooperatives.

Specialized cooperatives in pastoral areas should be established in line with local conditions and in diversified forms. On the basis of reality, the cooperatives should be developed by giving full play to the advantages of local endowments and local pillar industries, accelerating the growth of agricultural by-product processing industry, and extending their industrial channels. It is important to increase the added values of their products, so as to raise herdsmen's revenues while reducing ecological pressure on grasslands. At present, due to extensive and backward production modes in pastoral areas, prices of animal by-products are pretty low, which hinders the rapid improvement of herdsmen's income level. To address the issue, it is important to transform herdsmen's traditional ideas and encourage them to set up cooperatives, intermediary organizations or joint-

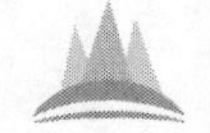

stock companies, in which herdsmen could become shareholders by selling their grassland use right, so as to enhance their market competitiveness and revenues.

The development of specialized cooperatives needs to be accelerated. Governments should increase financial support to the cooperatives. To solve the problems of insufficient start-up capital and development funds, it is important to expand financing channels and measures, such as low or discounted interest rates and agricultural policy-related loans. It is also necessary to encourage rural credit cooperatives and other forms of financial organizations to offer preferential policies toward agricultural credit projects. It is significant to cultivate specialized agents or households in pastoral areas, who might be the core elements for the development of a cooperative. It is advisable to develop circulation market for agricultural and livestock products so as to facilitate the growth of market-driven cooperatives for trading livestock products. It is also significant to introduce systems of registration and quality certification of agricultural products and to implement geographical branding strategy and standardized production of agricultural and animal husbandry products.

8.2.6 Strengthening the service system of husbandry production and promoting the management capacity for grassland sustainable development

It is advisable to formulate different animal husbandry development plans at various levels. With regard to the current industrial characteristics of agriculture and animal husbandry, it is important to clarify industrial development orientation and priorities for both short and long terms. Function, structure, scale and spatial distribution of modern husbandry service industry needs to be comprehensively designated through formulating scientific and reasonable long-term development plan. It is of great essential to improve and develop various types of service organizations and to further strengthen the development of the comprehensive service systems. Efforts should be made to establish and improve services networks of breeding on superior species, forage processing, animal disease prevention and control, grassland construction and technology promotion, so as to form a professional service network which is based upon national investment and participated by collective groups or herdsmen.

Modern and IT-based remote services of animal husbandry technology could be promoted through information network and service hotlines, such as the Internet, radio, telephone and other forms of telecommunications, which will help herdsmen and farmers to overcome difficulties and to resist the market risks and natural disasters. Aiming at serving the development of modern animal husbandry, it is necessary to establish a service system of scientific research, promotion, training, and information consultation. It is advisable to expand and strengthen service organizations to explore the marking operational modes. Efforts should be made to encourage animal husbandry and veterinary technicians to go to the production front to provide service to farmers and herdsmen through approaches such as technological contracting and shares. While offering services to farmers, technicians themselves would enhance their levels of techniques and revenues. In this way, modern animal husbandry with socialized service system in various forms and complete functions will be gradually formed.

A comprehensive and integrated service system for animal husbandry will be built up by taking advantages of local resources and with the animal products processing industry as a driving force. An integrated service network needs to be implemented, which will pave the way to a full range of animal husbandry. It would be necessary to explore the modes of running service organization with investment from different channels, such as governments, enterprises and individuals. Governments at different levels should encourage banks and other financial institutions to offer favorite support to cooperatives, such as low or discounted interest rate or even free loans in order to reduce the construction costs of the service organization as far as possible.

Chapter 9

Recommendations for Policy Design on Sustainable Development Management in Western Grasslands

9.1 Sustainable Development Policy-making in Western Grasslands

9.1.1 Key issues concerning policy-making of grassland management

In recent years, many countries have experienced grassland ecological degradation and sandstorm hazards. Taking the United States as an example, due to the adjustment and alteration in the western pastoral areas, "black storms" swept through the Western Great Plains since the 1930s, as a result, millions of hectares of farmland were abandoned and hundreds of millions of people were forced to be relocated. In the Former Soviet Union, a large area of land reclamation resulted in droughts and wind erosions, and about 17 million hectares of land were affected. In China, there exist different degrees of grassland degradation whether on the beautiful Inner Mongolia Prairie or the Qinghai-Tibet Plateau.

Grassland degradation refers to the degradation of the grassland ecosystem.

Its consequences are manifested in various aspects, among which the change of grassland vegetation is most direct and obvious. On severely degraded grasslands, the height and coverage of its bryophyte community would be remarkably decreased. Another manifestation of vegetation alteration is the changes in composition of bryophyte community. Under the conditions of excessive biting of livestock, plants which are not resistant to grazing would significantly vanish while those that are resistant to grazing would be preserved. Consequently, the degraded grassland will be composed of vegetations with low palatability. This is also the reason why the degraded grasslands may be composed of the vegetations that are resistant to droughts and grazing.

The general manifestation of grassland degradation includes grassland recession, grassland yield reduction, the growth of toxic and hazardous plants and inferior grasses, wind erosion and desertification, soil erosion and salinization, rampant rodents and pests. Grassland degradation, desertification and salinization accelerate soil erosion and climate change on the grassland, forming a vicious circle. The literature review shows that scholars with different backgrounds proposed various reasons behind the decline of grassland resources and the deterioration of the ecological environment. To sum up, the main points of view are as follows: meteorologically speaking, climate change is the basic cause of grassland degradation, and droughts have the main impact on grassland ecological degradation. From the perspective of policy-making, the inconsistency between agricultural policy and environmental policy is the main cause of grassland degradation. Grain-oriented policies in pastoral areas are blamed for the grassland ecological degradation. Other causes include the inconsistency between grassland contract responsibility system with that of dividing livestock to households, free use of grassland resources, insufficient management and low-efficiency of regulations. Some scholars pointed out that the government has made mistakes in adopting measures such as on fencing, stalk-feeding, breeding improvement, and migration. The formulation of relevant policies did not take the diversity of different areas and economic conditions into consideration, which

plays an important role in degradation.

Droughts are the main contributor to large-scale worldwide grassland degradation. Due to global greenhouse effect, temperature rises and rainfall decreases, leading to frequent droughts. During the spring, in particular, high temperature, droughts and fierce winds have a significant impact on grassland degradation. Water shortage will not only affect the general growth and yields of forage but also can cause diseases and pests on grasslands.

Human activities and severe policy intervention are also blamed for grassland degradation. To a large extent, grassland degradation is caused by the loss of traditional knowledge, such as "fixed habitation leads to the imbalanced utilization of pasture","contracting and enclosure results in grassland fragmentation", and "farming and mining industry caused grassland desertification". Animal husbandry production in China has long been evaluated in the light of the growth rate or the number of livestock rather than the quality of livestock. This has resulted in the one-sided pursuit of quantity, leading to a rise in stocking capacity of grasslands and the unbalance between livestock and forage. The evolution in which over-grazing causes grassland degradation is a gradual process: an increase in the number of livestock per unit area will lead to an increase in feeding forage per unit area, which will results in the insufficiency of the seeds for forage grass re-growth as well as an increase in poisonous weeds and hogweeds. Unless being checked, the vicious deadlock will lead to the increase of livestock and decrease of feeding forage per unit area. This will aggravate the physical destruction of soil structure in pastoral areas on the one hand, and will limit the growth of herbage on the other. This vicious deadlock will result in the gradually barren of grasslands, and finally their desertification and degradation. In the northern part of China, over-grazing is regarded as the second significant factors of grassland desertification.

In addition, grassland reclamation will not only cause the desertification of exploited arable land but also the desertification of its surrounding grasslands. Farming will destroy grassland vegetation and loose grass soil, which makes bare loose sandy soil to be very vulnerable due to wind erosion. When there is a wind, loosing sand becomes the source of the sand dunes. Too much reclamation would reduce the grassland area and increase the loading amount of livestock in pastoral areas. It would also result in vegetation degradation. The chain reaction would finally turn reclaimed lands into sandy areas.

Regarding the mechanism underlying grassland degradation, Ren Jizhou pointed out that the degradation of grasslands is due to the coupling between the animal production system and the plant production system, as well as the systematic discordance in the process of system coupling. That is the time discordance, spatial discordance and discordance among species during the system operation. The first is the time discordance. Yearly fluctuation of energy dynamics produced by animals in the natural state is pretty weak while those produced by vegetation is pretty strong. During the peak season of vegetation production, the profitability of vegetation organics would not only waste the grassland resources, but also lead to the inadequate utilization of grasslands. A lack of grazing will also deteriorate the composition of grassland vegetations and cause soil erosion, resulting in the damage to grassland resources. The second is spatial discordance. It has two meanings. One is spatial discordance among regions, that is the difference of livestock stocking per unit area of grassland among different regions. The other is spatial discordance between different seasons. During cold seasons smaller areas would undertake a much longer grazing interval with a heavier loading capacity while a much larger area will be devoted to a much shorter grazing interval during warm seasons with a lighter loading capacity. Finally, for the discordance among species, proper redistribution of ecological field and ecological niche is the key concern. With regard to the botanical composition of grasslands, well structured natural community or reasonable composition of different species would create a good ecological field to raise the yield of herbage, improve the forage quality and hinder diseases and pests of vegetation to keep fertility and prolong the service life of grassland. From the above three aspects, time discordance plays the leading and fundamental role,

which is due to the disparity on production systems between vegetations and animals.

In short, grassland degradation has become more and more severe due to the global climate change, the decrease of precipitation and the increase of population in pastoral areas. In the arid and semi-arid regions, the policy-adjustment of grassland management and utilization in different regions should be based on the local natural and climatic characteristics, and should not simply replicate the experience and methods of other regions. A more feasible management strategy suiting local conditions needs to be formulated by synthesizing the traditional experience and knowledge of the locals to address climate change and precipitation fluctuations. .

9.1.2 Fundamental basis for policy−making on grassland management

Facing large area of grassland degradation, the strict implementation of carrying capacity management became an important means of grassland governance in the 1990s. Although governments devoted a large amount of money, tangible capital and manpower to supervising and managing over-grazing, their efforts did not achieve expected results. The reason is that policymakers thought sustainable grassland management could easily be performed when keeping the number of livestock raised by herdsmen within the range of carrying capacity of the grasslands. The theory over-simplified and standardized the relationship among "herdsmen-livestock-grassland".

As a matter of fact, grassland is not the statistical understanding of vegetation coverage rate and the carrying capacity is not valued constantly but a dynamic variable at the micro-level. Characteristics of arid and semi-arid ecological system are of highly variable that can not be predicted. "Carrying capacity" and "loading capacity", which are two main grassland management tools supported by traditional management theory, are facing some inadaptability in arid and semi-arid areas. Scholars at home and abroad have conducted several researches on loading capacity of grasslands. Most of previous studies regard the utilization rate of forage as 50% when calculating loading capacity. However, in the majority of arid and semi-arid areas, due to the vulnerability and uncertainty of natural ecological conditions, the grasslands, where vegetation coverage have been destructed by grazing are prone to degradation and desertification. Many research results have rejected the theory regarding the utilization rate of forage as 50% during the calculation of loading capacity. According to the research results from the U.S. Bureau of Land Management, although grazing intensity of many places is in the permissible range of loading capacity, the grassland is still degrading, and even some areas with rare grazing has also faced some damage to vegetation community.

At present, many government officials or pastoral management personnel in arid and semi-arid regions focus upon the protection and recovery of perennial woody plants, striving to make their growth to an appropriate or optimal state so as to have the maximum amount of growth above the ground. However, species of short plants, annual plants and biennial plants and their values are ignored to a large extent. As a matter of fact, many annual and biennial plants survive to be the underground potential biomass as seed banks, whose sustainability and utilization value is of great significance to grasslands. In fact, some botanists and ecologists have figured out that due to the reproduction of a large number of annual or biennial vegetation, grazing played a minor role in grassland degradation process although the number of grazing livestock exceeded the theoretical loading capacity with the abundant precipitation.

In arid and semi-arid areas, ecosystem has a strong self-recovery ability, moderate grazing can not only promote the vegetation growth, especially the grass tiller, but also improve plant protein synthesis, the palatability of vegetation and its yield. Due to the impetus on vegetation re-production, grassland productivity would increase more than doubled in two weeks. However, with the increment of grazing intensity, grassland productivity decreased rapidly.

Experimental results show that in the pastoral areas that suffered from

severe degradation, grazing-banning should be implemented the whole year around, which would be beneficial to the vegetation recovery and the improvement indexes should include: yield, height, coverage, and diversity of grassland for some occasions. Measures on grazing-banning in the short run would achieve grassland recovery to some extent, but in the long run, the grassland would be degraded without the grazing livestock. Recovery of grassland after grazing-banning would go through a parabola. Degradation will follow up with a short while of recovery. Moderate intake is the important requirement for grassland restoration and the impact on grassland after grazing-banning is negative. Moreover, current experiments only take the harvested grass yield into calculation, however, in the non-grazing areas, besides the harvested grass, livestock intake has not been taken into calculation. Therefore, the current research is somehow incomplete.

In addition, the wide distribution of grassland resources in China faces an obvious seasonal imbalance, which is specifically manifested in the seasonal imbalance in pastoral area, in forage nutrition, in forage growth, and in the loading capacity of grassland. Due to the seasonal imbalance in grassland resources, livestock grazing depends highly upon the natural conditions, where lifecycle of livestock features "being strong in summer, fat in fall, thin in winter and death in spring". This is the common phenomena in most of the pastoral areas, which hinders the development of local economy and animal husbandry.

There is no denying that the carrying capacity of the grassland certainly has a limit. Beyond the threshold, the grassland will be degraded due to excessive consumption. Management depending on carrying capacity is a big step forward in grassland management policy-making, which contributes a lot to the sustainable development of grasslands. However, there is a constraint in the spatial dimension for the management depending on carrying capacity. On one hand, calculation of carrying capacity is affected by grassland resource, precipitation, natural hazards and other factors. Related studies have shown that carrying capacity management per household confronts a variety of problems that cannot be managed and implemented effectively. On the other hand, in large areas, carrying capacity management is to control the amount of livestock in a relatively stable system. Since livestock can be adapt to the forage yield with the spatial alteration of precipitation. This can effectively avoid the situation where sparse vegetation undertakes relatively more livestock grazing, and ensure that areas with abundant vegetations to be fully utilized, which would finally come to the livestock-forage balance.

9.1.3 Institution establishment for grassland management policies

Grassland-livestock double contract responsibility system (double contracting) refers to a policy allowing the contractual operation of collectively-owned pastures and livestock. This system has two components. One is the livestock contract system featuring "livestock pricing contract and cash retained" and "sold after pricing to herdsmen and amortized repayment". The other is pasture contract system. Under the policy, the responsibilities on grassland operation, management and construction would be contracted to households and the contractual interval is no less than 20 years. The policy is aimed at activating the initiative of herdsmen to achieve the double-goals on grassland protection and livestock feeding. In the early 1980s, a widespread implementation of the family contract responsibility system in rural areas proved that this policy scheme could greatly mobilize the enthusiasm of herdsmen and promote the development of agricultural production. Under this scenario, people thought the grassland-livestock double contract responsibility system could specify the responsibilities, rights and interests of herdsmen by contractual operation of livestock and grasslands. Herdsmen are entitled to the residual claim of operating income.

Carrying capacity control is the primary means of the double contracting system to achieve its goal of grassland protection. By controlling the quantity of livestock, a dynamic livestock-forage equilibrium can be achieved in some regions and during certain period of time. Grassland management model in

this double contracting system is actually the grassland succession model. Grazing density affects the process of vegetation succession by a certain ratio, and vegetation structure also confronts continuous alteration accordingly. The lower the grazing rate is, the better the condition of the grassland will be, and vice versa. However, the double contracting system is not a single policy. It is a comprehensive policy system for grassland management and livestock development, composing of a series of policy measures. It not only includes the alteration on property rights of livestock and grassland, but also a series of following reforms in animal husbandry and grassland management, including settlement for animal husbandry, drylot feeding, artificial forage plantation and carrying capacity management of grasslands.

The double contracting system is of positive significance to the development of animal husbandry during a certain time interval. First of all, from the perspective of animal raising pattern, this policy system overcomes the weakness of production means that depends upon natural conditions. Transformation from free grazing to intensive management on drylot feeding enhances the ability of grasslands to resist natural risks. Secondly, housing construction for livestock-raising and forage harvest during winter enhance the production capacity to some extent, which reduces the mortality and improves the survival rate of livestock in winter and spring. It can also somehow alleviate the productive fluctuation of "strong in summer, fat in fall, thin in winter and death in spring". Thirdly, contracting pastures to households basically put an end to irrational grazing and disordered competition. Herdsmen would focus on improving livestock slaughter rate and commodity rate to raise the household revenues rather than expanding the number of livestock.

Nevertheless, the implementation of the double contracting system would exert some negative impacts on grassland ecology, living standards and production of herdsmen and nomadic culture. Firstly, nomadic grazing system is thoroughly replaced by settled grazing. Pastures are divided into small patches by fencing, which limits the migration of wild animals. Livestock are limited within a fixed area of grassland grazing. Continuous grazing in the same area would accelerate the grassland degradation. Grassland fencing would lead to the monotonous vegetation structure and reduce the biodiversity of the ecosystem. It is not conducive to biodiversity evolution of natural grasslands, relative stability of the genetic resources and stable evolution of grassland. Secondly, livestock-forage balance management failed to reduce the utilization of grasslands. Instead, it stimulated the herdsmen to transform natural grassland ecosystem into artificial grassland ecosystem. Although this would improve grass yields, it enhances the intensity of grassland utilization, posing threat on the grassland ecological system. From the point of view of animal husbandry, perennial grassland resources are strongly resistant to livestock intake and trampling that will soon be able to regenerate and recover, so the carrying capacity of perennial grassland resources is far higher than that of annuals. Thirdly, management difficulties are huge and their costs are pretty high. Dynamic management and inspection on grassland and livestock should be conducted among millions of herdsmen, who are scattered and settled in the vast pastoral areas. However, costs for that would be difficult to predict.

9.2 Policy Design on Sustainable Development Management in Western Grasslands

9.2.1 Clarifying policy objects and improving policy relevance

1. Policy objects

Grassland ecological subsidy policy in essence is to adjust the interests of the relevant parties, so the reasonable definition of stakeholders has important significance for the policy improvement and implementation. In terms of the protection and utilization of ecological environment, stakeholders can be divided into beneficiaries, the impaired and polluters. In order to balance the interests of the stakeholders, principles of implementing ecological reward-compensation

policy should be: the beneficiaries pay for the benefits they gain; the impaired are compensated and the polluters are charged for the pollution costs. The interests of beneficiaries, being contemporary or future beneficiaries, can be represented by the government, as the government stands for the interests of all the people. For example, beneficiaries from grazing-banning could be herdsmen's descendants. The impaired parties include those impaired directly and indirectly. For instance, herdsmen who were forced to migrate for the ecological environment recovery of grasslands could be regarded as the direct impaired party; and those who have to reduce the quantity of livestock to maintain the livestock-forage balance could be regarded as the indirect impaired party since they have suffered from financial loss due to the decrease in the quantity of livestock. As to polluters, according to the existing policy, their major behavior is overloading. However, the eco-service function of some parts of the grassland ecosystem has been destroyed by mineral resource exploitation and ecological tourism in the region. Those polluting enterprises have not paid for their interruptive behaviors, particularly the damage they did to the local households whose grasslands were not occupied by the enterprises. Therefore, a system of taxation on ecological compensation should be set up. Taxation for ecological compensation should be levied upon the behavior entities that interrupt the ecological environment of grasslands. This part of taxation revenue should be used for the special purpose of ecological protection and construction.

2. Policy relevance

The implementation of grassland ecological compensation policy is a major measure to protect the ecological environment and promote the sustainable development of animal husbandry. To be effective in implementation, the policies should be consistent and stable during a given period. Although the field study has found problems with the implementation of ecological compensation policies, such as the way to coordinate the relationships among environmental protection, development on animal husbandry development and herdsmen's livelihood, this should not be the reason to renounce the continuous implementation of the ecological compensation policy. Therefore, we should enhance the consistency of this policy and make more efforts to balance environmental protection with livelihood of herdsmen by using market mechanism to solve the environmental problems to the utmost extent in place of administrative means.

9.2.2 Legislation on grassland ecological policies

China has issued several laws on grassland ecological environment protection, including *The Environmental Protection Law of the People's Republic of China*, *Law of the People's Republic of China on Water and Soil Conservation* and *Law of the People's Republic of China on Grasslands*. However, compared with laws and regulations on ecological protection in developed countries, China has not formulated a normative system of ecological compensation yet. China's ecological compensation legislation lags far behind its management practice. Therefore when dealing with the ecological and environmental problem, it always lacks effective institutional support. Due to the lack of clear definition and specification on related parties of interests, the compensation actions sometimes are arbitrary and become an approach to deal with emergencies. When conditions are permitted, rules on ecological compensation need to be enacted through legislation. Ecological compensation should be taken into the scope of public finance in multiple forms, such as direct government subsidies, projects for ecological protection and horizontal economic compensation among regions.

9.2.3 Auxiliary policy for financial support

First of all, intensity of the central financial investment should be enhanced. It is necessary to increase the reward-compensation level by taking livestock-forage balance and living costs of herdsmen into consideration. In addition, more financial support should be increased continuously for migrating herdsmen's skill-training, children's education, social security and other aspects. Financial

and economic structure in pastoral areas needs to be adjusted so as to speed up the reform on rural cooperative organizations. With support from central and local governments, property rights of credit cooperatives should be clearly defined. A sound internal governance structure needs to be established by encouraging commercial banks to participate into the economic development of pastoral areas to guide the development of private finance sector. Then, banks should extend the terms of loans and shorten their approval duration in light of production reality and the operation cycle of animal husbandry. Due to the scattered settlement of herdsmen and their diversified loan demand, banks should consider the loan purposes according to different levels of economic development and formulate flexible loan policies. If conditions are permitted, loan period could be appropriately extended. Finally, it would be necessary to allow herdsmen to mortgage their grassland contract rights, which would be one of their sources to obtain loans. The grassland is the most precious means of production for herdsmen. If grassland resources could be mortgaged, it would play an important role for the improvement of herdsmen's living and production level through the access to credit funds.

9.2.4 Improvement on social security system in pastoral areas

In order to realize the medical care and services for the aged, a social security system should be practically implemented in pastoral areas. First, legislation on social security in pastoral areas needs to be speeded up to provide legal support for the establishment of a social security system. It is important to clarify the characteristics of social security, responsibilities of governments and enterprises, and rights and obligations of herdsmen. Then, a rural cooperative medical system (CMS) needs to be further improved on the basis of "voluntary-participation principle". Specific standards of minimum payment and reimbursement need to be scientifically set up to ensure herdsmen can afford the medical care and services. Additionally, organization construction, supervision and management of pastoral areas need to be improved and strengthened.

Efficiency of social security work in pastoral areas needs to be improved by simplified administration. Finally, by strengthening the publicity of social security system in pastoral areas, sense of identity among herdsmen on social security system would be enhanced.

9.2.5 Implementation of comprehensive poverty−alleviation policy

Poverty in pastoral areas is not only an economic issue, but also an ecological problem. The national financial aid to the poor in the pastoral areas should be continuously increased with economic development and fiscal revenue growth. However, solutions to poverty in pastoral areas rely on the alteration of economic thinking to solve the problem. In the pastoral areas of Inner Mongolia, due to the backwardness of second and tertiary industries, educational level of herdsmen is limited. Animal husbandry is still the main industry in these areas. Governments should create conditions as far as possible for herdsmen to benefit from the process of modernization. It is significant to provide herdsmen with job opportunities in urban, industrial and mining enterprises especially after the introduction of grazing-banning policy. A large number of skill trainings needs to be arranged for herdsmen so that they could be smoothly transferred to the employment of the tertiary industry. In addition, relying on government's financial supports as the only source of funding for alleviating poverty is not realistic. It is of essential to vigorously promote private capital into the projects on poverty alleviation to encourage more NGOs to participate into the related projects. Finally, it should be pointed out that although there is certain linkage between grassland ecological compensation policy and poverty alleviation, the two are different things in essence. The purpose for grassland ecological reward-compensation policy is to protect ecological environment and to restore the damaged environment while the purpose for poverty alleviation is to shorten the gap between the rich and the poor to realize social justice. Implementation of grassland ecological reward-compensation policy is just one of the means

to alleviate poverty in pastoral areas. Also, in light of actual situations of the local and herdsmen production and livelihood, more efforts need to be devoted to support the infrastructure construction in pastoral areas. Financial support and technical assistance are needed to the construction on wells, fencing and greenhouses. This would help herdsmen overcome practical difficulties and reduce the burden in living and production which would finally realize the transformation on animal husbandry production mode.

9.2.6 Establishing a mechanism underlying policy-making with herdsmen's participation

Before formulating any policy regarding grasslands and pastoral areas, it is important to listen to the opinions from herdsmen. Herdsmen who have lived through the traditional life in pastoral areas with extensive operation experience are best qualified to speak on the management and operation of grasslands. During the field investigation of grassland ecological reward-compensation policy implementation, most herdsmen thought that the compensation standards are pretty low, and should be set to a reasonable level of 3 to 4 yuan. They said if the punishment is not sufficient, they would prefer accepting penalty to reducing livestock.

During the process of the coordinated development of grassland ecology, economy and society, it would not be desirable to formulate "one-way or top-down" policies without listening to opinions and suggestions from herdsmen. It is necessary to ask governments to listen to the voice from herdsmen through collaboration between governments and herdsmen. The common goal on environmental protection would reach the mutual understanding if the policies would be formulated with the herdsmen's opinions and experiences. A scientific and reasonable policy system would be formulated in accordance with the real situation of local compensation standards, which is a win-win approach to both achieve ecological environmental protection and economic sustainable development of pastoral areas.

References

Bai Y F, Han X G, Wu J G, et al. 2004. Ecosystem stability and compensatory effects in the Inner Mongolia grassland[J]. Nature, 431:181-184.

Bai Y F, Huang J H, Zheng S X, et al. 2014. Drivers and regulating mechanisms of grassland and desert ecosystem services [J]. Chinese Journal of Plant Ecology, 38 (2): 93-102.

Bureau of Animal Husbandry of Ministry of Agriculture, PRC. 1996. Grassland Resources of China [M].Beijing: China Science and Technology Press. (In Chinese)

Cai H, Zhang X F, Dong S K, et al. 2013. A participatory survey on status and local pastoralists' consciousness to animal husbandry in alpine grassland [J].Grassland and Turf, 33(2): 66-72. (In Chinese)

Cao Y J. 2013. Study on the Ecological Compensation of Grassland in Xilinguole Meng [D]. Inner Mongolia: Inner Mongolia University. (In Chinese)

Chadden A, Dowksza E, Turner L. 2004. Adaptive management for southern California grasslands [J].University of California, Santa Barbara: Donald Bren School of Environmental Science and Management.

Chen J, Luo D. 2009. Investigation on Grassland Ecological Management in China [M].Shanghai: Far East Publishing House. (In Chinese)

Cheng G D, Zhao C Y. 2008. An integrated study of ecological and hydrological processes in the inland river basin of the arid regions, China [J]. Advances in Earth Science, 23(10): 1005-1012.

Costanza R, D'Arge R, de Groot R, et al. 1997. The value of the world' s ecosystem services and natural capital [J]. Nature, 387(6630): 253-260.

Da L T, Naren G W. 2016. Reflections on the theory and system of animal husbandry in Inner Mongolia [J]. Northern Economy, 6: 32-35. (In Chinese)

Deng B. 2014. Sustainable Development on the System Carrying Capacity of Grassland Industrial Ecosystem [D]. Lanzhou: Gansu Agricultural University. (In Chinese)

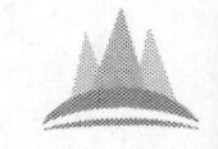

Du F C, Cao Q. 2012. Eco-husbandry: trends of animal husbandry economy in West China Prairies [J]. Journal of Original Ecological National Culture, 4(1): 118-127. (In Chinese)

Duan Q W, Chen B R, Zhang H B, et al. 2012. Discussion on the theory and policy of the development of grassland animal husbandry [J].Grassland and Turf, 32(6): 67-73. (In Chinese)

Dymond C C, Johnson E A. 2002. Mapping vegetation spatial patterns from modeled water, temperature and solar radiation gradients [J]. Photogrammetry and Remote Sensing, 57: 69-85.

Feng X Z, Li X T. 2013. Discussion on perfection of grassland law system in China [J]. Pratacultural Science, 30(1): 148-151. (In Chinese)

Feng Y, Wang W J, Liu J H, et al. 2013. Spatial and temporal variation of main impact factors and importance of wind break and sand-fixing function in Hulun Buir Grassland Eco-function Area[J]. Journal of Environmental Engineering Technology, 3(3): 220-230. (In Chinese)

Franklin J. 1995. Predictive vegetation mapping: geographic modeling of bio-spatial pattern in relation to environmental gradients [J]. Progress in Physical Geography, 19(4): 474-499.

Gao S Y, et al. 2008. Benefits of Beijing-Tianjin Sand Source Control Engineering [M]. Beijing: Science Press. (In Chinese)

Gong F, Chang Q. 2012. Research on the Construction and Application of the Government-Dominated Grassland Ecological Compensation Mechanism in China [M]. Beijing: Economic Science Press. (In Chinese)

Guo J, Lang X. 2005. Ecological and environmental status and measures to functional restoration of Western Grassland [J]. Acta Ecologiae Animalis Domastici, 26(5): 1-3. (In Chinese)

Ha H R. 2006. Study on Sustainable Development of Animal Husbandry Pasturing Areas in Inner Mongolia Xiwu Zhu Muqin Banner [D]. Beijing: Chinese Academy of Agricultural Sciences. (In Chinese)

Hao Y B, Wang Y F, Cui X Y. 2010. Drought stress reduces the carbon accumulation of the *Leymus chinensis* steppe in Inner Mongolia, China [J]. Chinese Journal of Plant Ecology, 24: 898-906.

Holechek J L, Pieper R D, Herbel C H. 2001. Range Management Principles and Practices [M]. 4th ed. New Jersey: Prentice Hall.

Hong D X. 2012. Eco-compensation Mechanisms of Ecological Environmental Construction of Grassland [M]. Beijing: Economy and Management Publishing Press. (In Chinese)

Hou X Y, Wan L Q, Gao H W. 2004. Research on keystones of grassland science in Western China [J]. Scientia Agricultura Sinica, 37(4): 558-565. (In Chinese)

Hou X Y, Yin Y T, et al. 2011. An overview and prospects for grassland adaptive management in China [J]. Acta Prataculturae Sinica, 20(2): 62-69. (In Chinese)

Janet F. 1995. Predictive vegetation mapping: geographic modeling of bio-spatial pattern in relation to environmental gradients [J]. Progress in Physical Geography, 19(4): 474-499.

Jiang L P, Qin Z H, Xie W, et al. 2007. Estimation of grassland ecosystem services value of China using remote sensing data [J]. Journal of Natural Recourses, 22(2): 161-170.

Lesica P, Kittelson P M. 2010. Precipitation and temperature are associated with advanced flowering phenology in semi-arid grassland [J]. Journal of Arid Environments, 74: 1013-1017.

Li B. 1997. The rangeland degradation in north China and its preventive strategy [J]. Scientia Agricultura Sinica, 30(6): 1-9.

Li J, Dai N N, Liu S Y. 2011. The summary of the settlement of nomadic pastoralists of Western Grassland [J]. Journal of Inner Mongolia University for Nationalities (Social Sciencies), 37(3): 7-10. (In Chinese)

Li W H, Zhang B, Xie G D. 2009. Research on ecosystem services in China: progress and perspective [J]. Journal of Natural Recourses, 24(1): 1-10.

Li Y B, Li W J. 2012. Why "balance of forage and livestock" system failed to reach sustainable grassland utilization [J]. Journal of China Agricultural University(Social Sciences Edition), 29(1): 124-131. (In Chinese)

Liebig M A, Gross J R, Kronberg S L, et al. 2010. Grazing management contributions to net global warming potential: a long-term evaluation in the Northern Great Plains [J]. Journal of Environmental Quality, 39: 799-809.

Liu X N, Sun J L, et al. 2008. A study on the community structure and plant diversity of alpine meadow under different degrees of degradation in the Eastern Qilian Mountains [J]. Acta Prataculturae Sinica, 17(4): 1-11. (In Chinese)

Liu X Y, Liang T G, Long R J, et al. 2009. Classification management mechanisms for grassland resources and sustainable development strategies in Northern China [J]. Acta Ecologica Sinica, 29(11): 5851-5859. (In Chinese)

Liu X Y, Long R J, Shang Z H. 2011. Evaluation method of ecological services function and their value for grassland ecosystems [J]. Acta Prataculturae Sinica, 20(1): 167-174. (In Chinese)

Lu Y, Zhuang Q, Zhou G, et al. 2009. Possible decline of the carbon sink in the Mongolian Plateau during the 21st century [J]. Environmental Research Letters, 4: 1-8.

Mattia T, O' Farrell P, Eduardo N, et al. 2014. Mapping ecological processes and ecosystem services for prioritizing restoration efforts in a semi-arid Mediterranean river basin [J]. Environmental Management, 53:1132-1145.

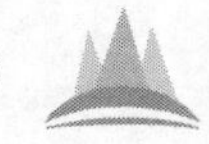

Meng H J, Teng Y Z, Liu Z L. 2006. On the problems and bewilderments faced by the sustainable effectiveness of investment in the grassland ecology construction [J]. Journal of Inner Mongolia Finance and Economics College, 1: 39-42. (In Chinese)

Oldeman, et al. 1981. World map of the human-induced soil degradation. An explanatory note. Wageningen: International soil Reference and Information Certer/Nairobi. Nairobi: United Nations Environment Programme.

Peng S Z, Zhao C Y, Bie Q. 2010. Land classification based on remote sensing data and analysis of spatial distribution in Zuli River Basin [J]. Remote Sensing Technology and Application, 25(3): 366-372.

Peng S Z, Zhao C Y. 2014. Mapping daily temperature and precipitation in the Qilian Mountains of Northwest China [J] . Journal of Moutain Science, 11(4): 896-905.

Piao S L, Fang J Y, Ciais P, et al. 2009. The carbon balance of terrestrial ecosystems in China [J]. Nature, 458:1009-1014.

Pu X P, Shi S L, Yang M. 2011. Ancient Chinese grassland protection regulation and its impact on modern grassland management [J]. Grassland and Turf, 31(5):85-90, 96. (In Chinese)

Ren J Z, Liang T G, Lin H L, et al. 2011. Study on grassland's responses to global climate change and its carbon sequestration potentials [J]. Acta Prataculturae Sinica, 20: 1-22. (In Chinese)

Schuman G E, Janzen H H, Herrick J E. 2002. Soil carbon dynamics and potential carbon sequestration by rangelands[J]. Environmental Pollution, 116:391-396.

Song H Y, et al. 2006. Research Report on the Development of Grassland Improvement and Pastoral Areas in China [M].Beijing: Beijing: China Financial and Economic Publishing House. (In Chinese)

Song L H, Tang X H. 2012. The foundation and path of development of grassland carbon sink economy in Inner Mongolia [J]. Chinese Journal of Grassland, 34(2): 1-7. (In Chinese)

Straton A. 2006. A complex systems approach to the value of ecological resources [J]. Journal of Ecological Economics, 56: 402-411.

Su D X. 2001. The management, perfection, and regional planning of grassland in west China [J].Journal of China Agricultural Resources and Regional Planning, 22(1): 14-15. (In Chinese)

Su Y B, Zhang F P, Feng Q, et al. 2015. Estimation of alpine grassland biomass and analysis if its spatial distribution characteristics in typical small watershed of Qilian Mountain [J]. Journal of Shanxi Normal University (Natural Science), 43(2): 79-84.

Wang C. 2013. The Impact of Vegetation Chante on Rainfall-runoff Process in Tianlaochi Catchmont in the Heihe River Basin [D]. Lanzhou: Lanzhou University.

Wang J B, Zhang D C, Tian Q. 2013. Evaluating herders' knowledge of grassland policy in arid and semi-arid pastoral areas in central and western Gansu [J]. Acta Agrectir Sinica, 21(1):11-17. (In Chinese)

Wang J Y, Chang X X, Ge S L, et al. 2001. Vertical distribution of the vegetation and water and heat conditions of Qilian Mountain (northern slope) [J]. Journal of Northwest Forestry University, 16 (Supp.) : 1-3.

Wang M J, Ma C S. 1994. A study on methods of estimating the carrying capacity of grassland [J]. Grassland of China, 16(5): 19-22.

Wang Q L, Wang C T, Du Y G, et al. 2008. Grazing impact on soil microbial biomass carbon and relationships with soil environment in alpine Kobresia meadow [J]. Acta Prataculturae Sinica, 17(2):39-46. (In Chinese)

Wang S J, Jiao S T, Wang S X. 2010. Ecological security construction and grassland eco-tourism development [J].Chinese Journal of Grassland, 32(6): 101-104,109. (In Chinese)

Wang S P. 2006. Theory and practice on sustainable use of China’ s natural grassland: a study on the development strategy of grassland-livestock and grassland-agriculture [J]. Acta Agrestia Sinica, 14(2): 188-192. (In Chinese)

Wang S, Wilkes A, Zhang Z, et al. 2011. Management and land use change effects on soil carbon in northern China’ s grasslands: a synthesis [J]. Agriculture, Ecosystems and Environment, 142:329-340.

Wang W Y, Wang Q J, Wang G. 2006. Effects of land degradation and rehabilitation on soil carbon and nitrogen content on alpine Kobersia meadow [J]. Ecology and Environment, 15:362-366. (In Chinese)

Wang X Y. 2009. Interactive community management: an experiment of grassland management through democratic negotiations in Pifang Villagers Group in Hexigten Qi [J].Open Times, 4: 36-49. (In Chinese)

Wang Z B, Yu J, Liu X W. 2009. Research on the relationship between ecosystem services and ecological compensation [J]. China Population Resource and Environment, 19(6): 17-22.

Wong N, Morgan J. 2007. Review of Grassland Management in South-eastern Australia [M]. Melbourne: Parks Victoria.

Wu Y S, Ma W L, Li H, et al. 2010. Seasonal variations of soil organic carbon and microbial biomass carbon in degraded desert steppes of Inner Mongolia [J]. Chinese Journal of Applied Ecology, 21: 312-316. (In Chinese)

Xie M H, Ta N. 2013. A case study on China’s four prairies: variable weight comprehensive evaluation of triple capacities of prairie sustainable development [J].Resources and Industries, 15(4): 118-124. (In Chinese)

Xu Z, Zhao C, Feng Z. 2012. Species distribution models to estimate the deforested area of Picea crassifolia in arid region recently protected: Qilian Mts. National Natural Reserve

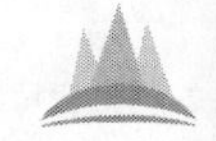

(China) [J] . Polish Journal of Ecology, 60(3): 515-524.

Yang B J, Ma Y X, Li B. 2010. Protection and construction of grassland in China: investigation in the Inner Mongolia [J]. China Development, 10(2):1-5. (In Chinese)

Yang B J, Zhang L J, Wang X B, et al. 2011. The construction and management of natural grassland reserves: survey on the natural grassland reserves of Xinjiang Uygur Autonomous Region [J]. China Development, 11(1): 1-5. (In Chinese)

Yang L H, Ouyang Y. 2012. Tactical man and the tactical decision-making model: agent-based simulations of grassland management [J]. Chinese Public Administration, 8: 106-110. (In Chinese)

Yang L H. 2011. Multi-collaborative governance: a game theoretical model of grassland management and empirical studies [J]. Chinese Public Administration, 4: 119-124. (In Chinese)

Yang L. 2010. Dilemma of grassland governance in China: from "Tragedy of Commons" to "Fencing Dilemma" [J]. China Soft Science, 1: 10-17. (In Chinese)

Yang Y H, Fang J Y, Tang Y H, et al. 2008. Storage, patterns and controls of soil organic carbon in the Tibetan grasslands [J]. Global Change Biology, 14:1592-1599.

Yu G, Lu C X, Xie G D, et al. 2006. Water holding capacity of grassland ecosystem and their economic valuation in Qinghai-Tibetan Plateau based on RS and GIS [J]. Journal of Mountain Science, 24(4): 498-503.

Yuan C X. 2014. Study on the Driving Mechanism of the Oasis Dynamic Changes in the Heihe River Basin [D]. Lanzhou: Lanzhou University.

Zhang H, Shen W S, Wang Y S, et al. 2005. Study on grassland grazing capacity in the Heihe River Basin [J]. Journal of Natural Recourses, 20(4): 514-521.

Zhang S Q, Yan W G. 2006. Problems of grassland ecosystems and their countermeasures in western China [J]. Acta Prataculturae Sinica, 15(5): 11-18. (In Chinese)

Zhang Z Q, Xu Z M, Wang J et al. 2001. Value of the ecosystem services in the Heihe River Basin [J]. Journal of Glaciology and Geocryology, 23(4): 360-366. (In Chinese)

Zhao C Y, Bie Q, Peng H H. 2010. Analysis of the niche space of Picea crassifolia on the northern slope of Qilian Mountains [J]. Acta Geographica Sinica, 65(1): 113-121.

Zhao C Y, Feng Z D, Nan Z R, et al. 2007. Modelling of potential vegetation in Zulihe River Watershed of the West-central Loess Plateau [J], Acta Geographica Sinica, 62(1): 52-61.

Zhao C Y, Nan Z R, Cheng G D, et al. 2006. GIS-assisted modelling of the spatial distribution of Qinghai spruce (Picea crassifolia) in the Qilian Mountains, northwestern China based on biophysical parameters [J]. Ecological Modelling, 191(3): 487-500.

Zhao C Y, Nan Z R, Cheng G D. 2005. Methods for modelling of temporal and spatial distribution of air temperature at landscape scale in the southern Qilian mountains, China [J]. Ecological Modelling, 189(1): 209-220.

Zhou L, Dong X Y. 2013. The Systematic logic concerning the problems of pastoral area, animal husbandry and herdsman—research on pasture management and property rights system transition in China [J]. Journal of China Agricultural University (Social Sciences Edition), 30(2): 94-107. (In Chinese)

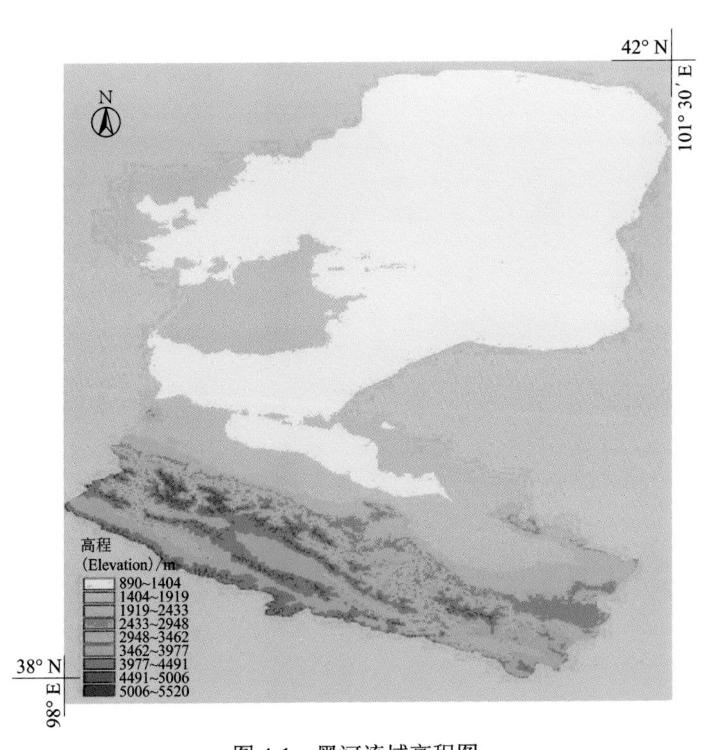

图 4-1　黑河流域高程图

Figure 4-1　DEM of the Heihe River Basin

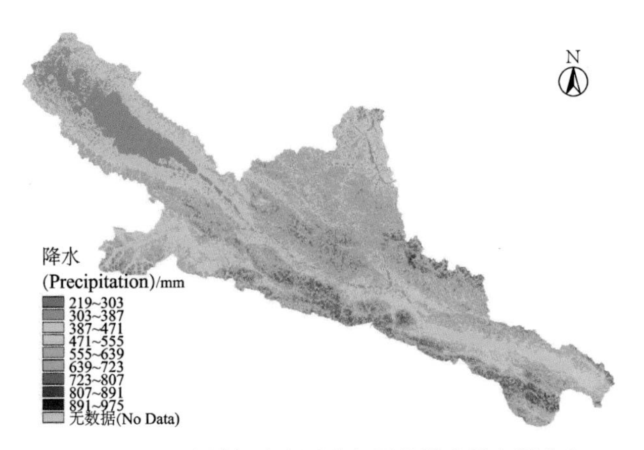

图 4-3　黑河流域祁连山区多年平均降水量空间分布

Figure 4-5　The spatial distribution of annual mean precipitation in the upper reach of Heihe River

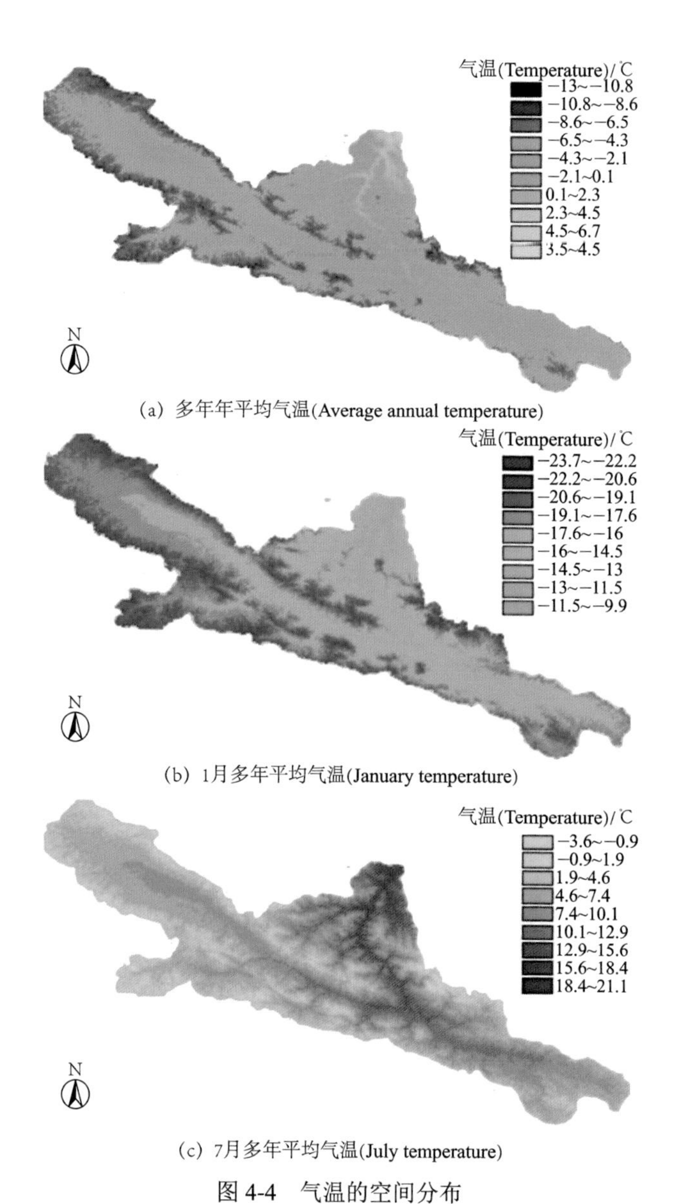

(a) 多年年平均气温(Average annual temperature)

(b) 1月多年平均气温(January temperature)

(c) 7月多年平均气温(July temperature)

图 4-4 气温的空间分布

Figure 4-6 The spatial distribution of temperature in different time

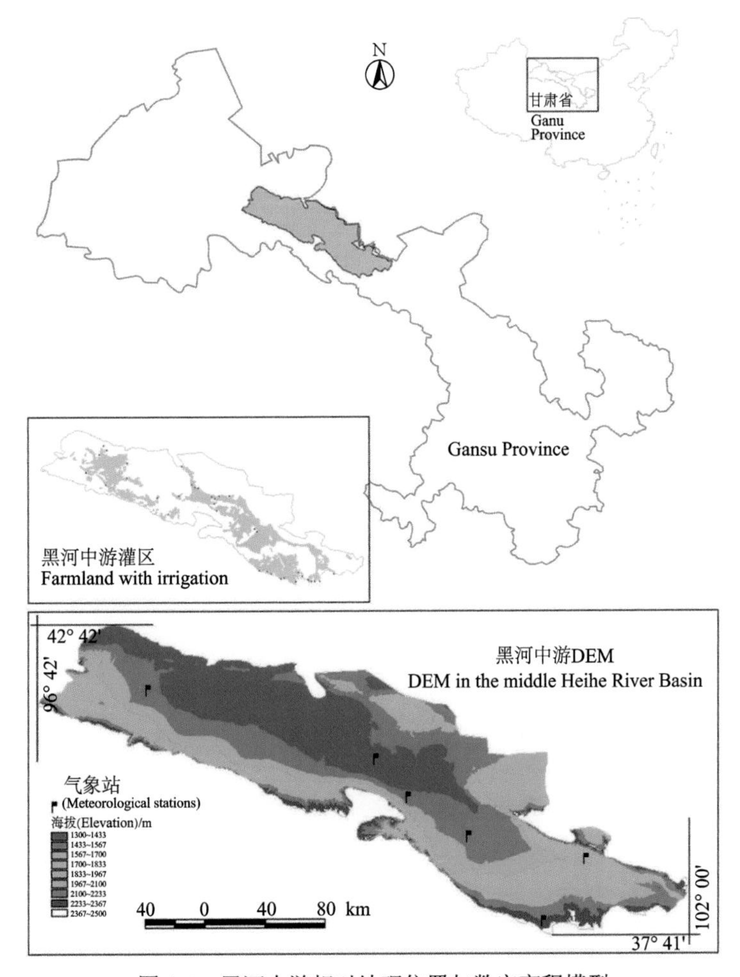

图 4-5　黑河中游相对地理位置与数字高程模型

Figure 4-7　Location and DEM of the middle reaches of the Heihe River basin

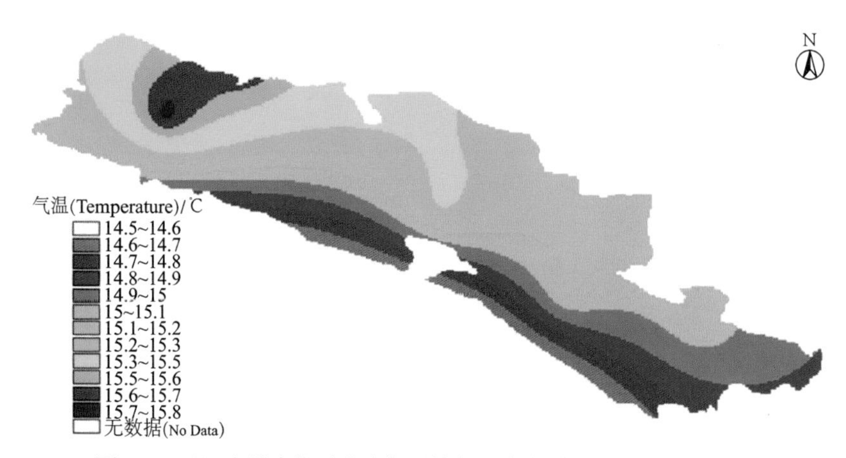

图 4-6　黑河中游生长季节多年平均气温分布图（1960 ～ 2014 年）

Figure 4-8　Distribution of annual air temperature during growing season in the middle reaches of the Heihe River Basin (from 1960 to 2014)

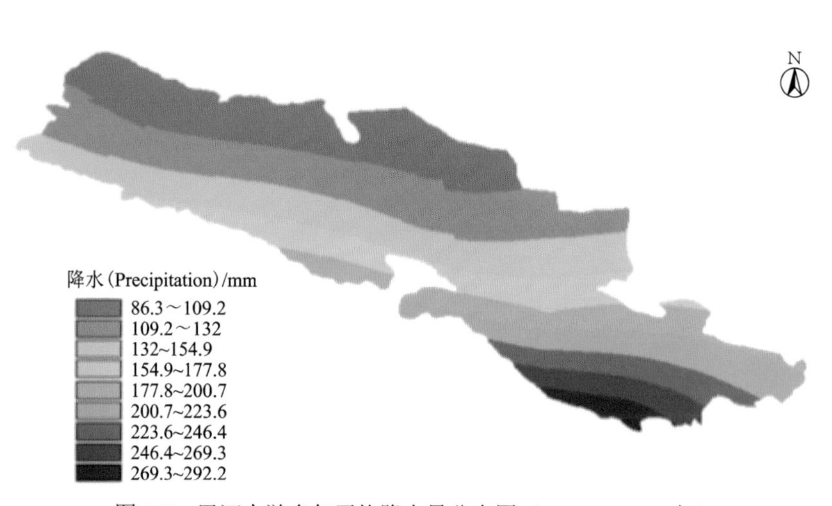

图 4-7　黑河中游多年平均降水量分布图（1960 ～ 2014 年）

Figure 4-9　Distribution of annual mean precipitation in the middle reaches of the Heihe River Basin (from 1960 to 2014)

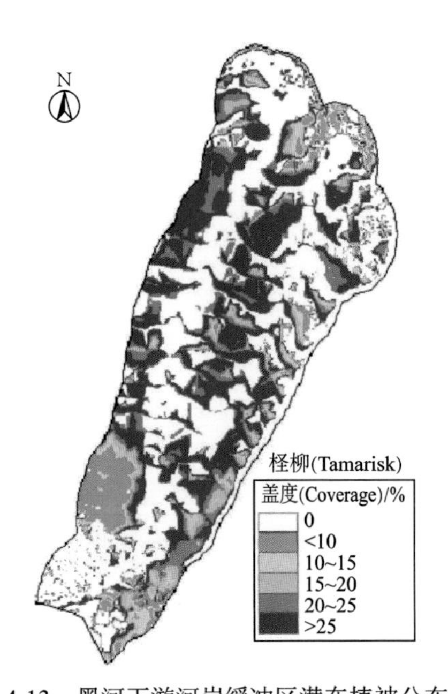

图 4-13　黑河下游河岸缓冲区潜在植被分布

Figure 4-15　Spatial distribution of grassland in the river's buffer in lower reaches of the Heihe River Basin

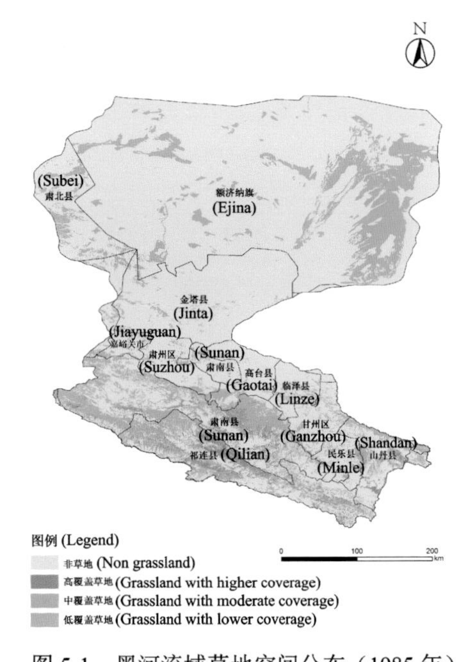

图 5-1　黑河流域草地空间分布（1985 年）

Figure 5-1　The distribution of grassland in the Heihe River Basin in 1985

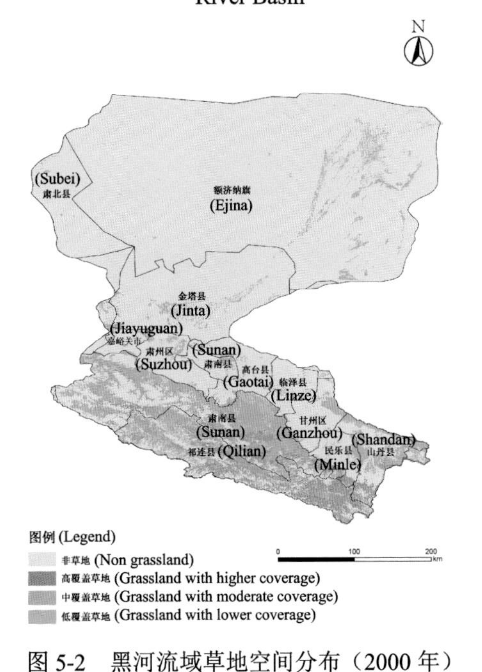

图 5-2　黑河流域草地空间分布（2000 年）

Figure 5-2　The distribution of grassland in the Heihe River Basin in 2000

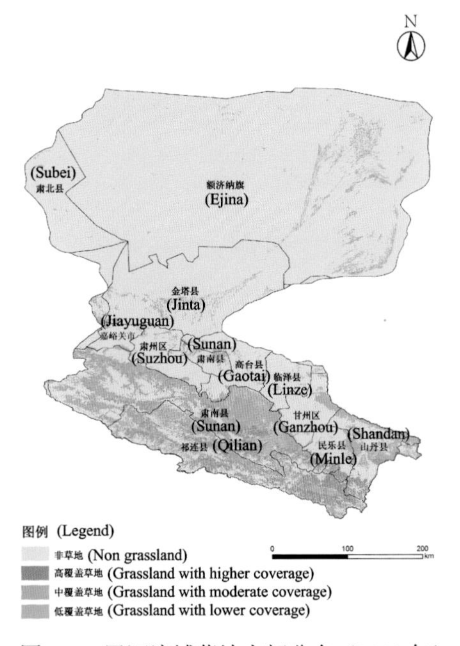

图 5-3　黑河流域草地空间分布（2013 年）

Figure 5-3　The distribution of grassland in the Heihe River Basin in 2013

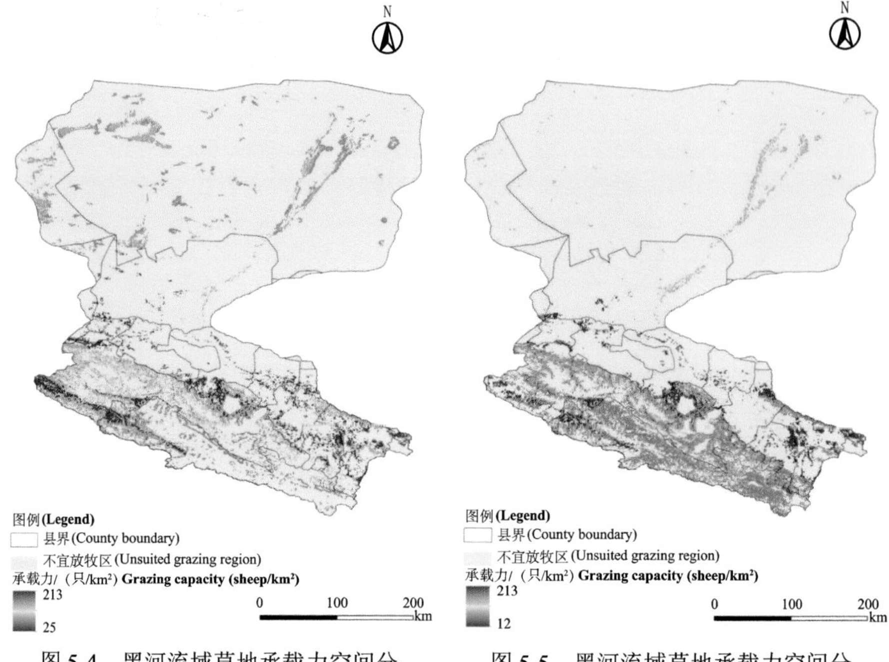

图 5-4　黑河流域草地承载力空间分布（1985 年）

Figure 5-4　Spatial distribution of grassland carrying capacity in the Heihe River Basin (1985)

图 5-5　黑河流域草地承载力空间分布（2000 年）

Figure 5-5　Spatial distribution of grassland carrying capacity in the Heihe River Basin (2000)

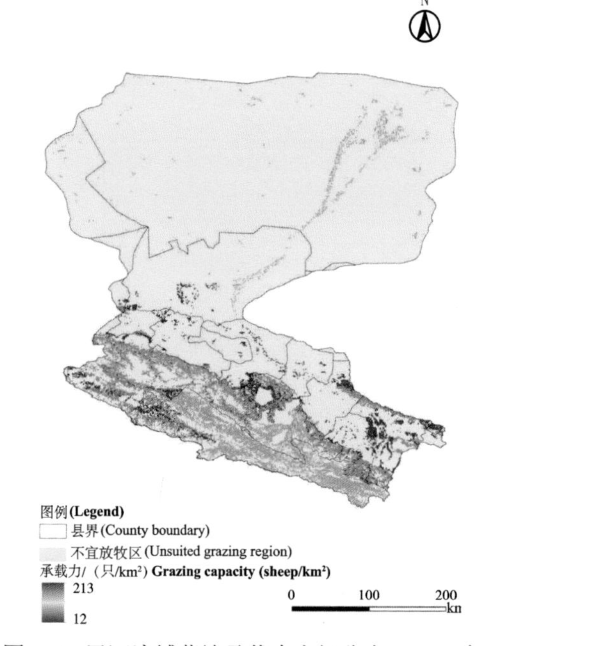

图 5-6　黑河流域草地承载力空间分布（2013 年）

Figure 5-6　Spatial distribution of grassland carrying capacity in the Heihe River Basin (2013)